Bid Up Now!

Auction Marts, Fairs & Droving
Over Two Centuries
In Yorkshire, Lancashire and Cheshire

by John A Thomson F.I.A. Scot.

CONTENTS

Livestock Auctioneers Market Committee for England and Wales 1971
(front row) T.R. Hill, J. Muir Watt, J.A. Thomson, (author), H. Lacy Scott, R.F.H. Stephenson, E.C. Dee, Richard Turner

INTRODUCTION

From about 1850 livestock auctions brought much needed financial stability to the breeders and feeders of the North of England by providing a means of ascertaining the value of their livestock and guaranteeing payment of the full price. This development had been delayed and hindered by a tax of five per cent, which had been levied on all auction sales of farm stock which were held other than on the seller`s farm. However, in 1846 common sense had prevailed and the tax was abolished. This, coupled with the introduction and rapid development of railways, provided an opportunity for enterprising auctioneers and farmers to open the first few of what would become one hundred and sixty marts in the counties of Yorkshire, Lancashire and Cheshire; sometimes with several in one town or with the auctioneers often sharing a council-provided mart. In Hull a record number of over twenty firms would operate on the same day and such was the pattern in Lancashire and Yorkshire and, to a lesser extent, in Cheshire until the early 1980`s when the rate of closures would increase and we would see the loss of more than three quarters of these marts. It is sad to think that in Cheshire only one mart remains where once there were over twenty-five, and in the three counties of Yorkshire, Lancashire and Cheshire no more than twenty continue to hold weekly sales.

The newly established marts were an immediate success. Farmers flocked to support them, not only to attend the sale but to meet their corn merchant, their seedsman, their implement merchant and their friends. Getting on their better boots, they made it the focus of their week. Their wives looked forward to it as they got a lift with the shopping and, if their husband came home a little cheery, perhaps it added to the pleasure.

In this book, I have continued my brief histories of British auction marts into Lancashire, Cheshire and Yorkshire at the suggestion of a number of those who have been good enough to read my previous books on the marts of Scotland, “Ring of Memories”, and the marts of the English northern counties, “Farmers Went to Market”. I apologise to these regular readers who will notice that there has, in the early chapters, been an amount of unavoidable duplication in some of the background historical information which is so material to the full story.

In my previous books I had a good geographical grasp of the roads into every mart and the countryside surrounding them, but in the area covered by this volume my knowledge has been more limited and most of the marts referred to were never visited by me or my family. I have inevitably found this to be something of a problem as I frequently found myself asking "where exactly is this place?". However, I have had much help in this regard and hope that I have not blundered too often.

Preparing these books, I have realised that I have perhaps come to this some twenty years too late as many of our older auctioneering friends have left the rostrum for a higher calling. However, the younger auctioneers have been more than willing to give up their time to talk to me and to assist me with information and I have been tremendously impressed by the quality of men and women currently in our profession. Dates of events have sometimes been difficult to recall or substantiate so many should be considered as being approximate. The book is a mix of history and reminiscences and I hope that those who are seeking only one of these will forgive the other.

Researching has confirmed me in my life-long belief of the long-standing importance of the marts in getting good, healthy, fresh meat from the farm to the dinner table and in ensuring a true open market for the farmers` product. In this regard, they remain as important as ever and fully justify continued and increased support from the producers whom they serve.

As you read this, reminiscences and corrections may spring to mind and I will be delighted to hear them from you in case a second edition is, at some time, produced by my long-suffering family of auctioneers.

HELPERS

There have been so many people who have kindly allowed me to take up their time and share their recollections. My memory has, however, deteriorated a little and, in order to try not to offend anyone, I have decided not to attempt to list all who helped. However I must mention a few people who have particularly made it possible for me to produce this book and to whom special thanks are due.

Stephen Walker, who trained as an auctioneer with me at Annan and now manages the successful mart in Leyburn, has given me his time and knowledge from start to finish and has now produced a good history of his own mart.

Mervyn Lister has an unrivalled archive and knowledge of livestock auctioneering, which he has shared most generously with me. I strongly recommend his book, "On This Day at the Auction Mart" (Otley and Bingley Marts).

Other must-have books are "Winds of Change" by Peter Atkinson of Driffield and "The Fall of the Hammer" by Brian Pennington of Beeston Castle.

George Chapman of Masham let me use his archives and shared his amazing knowledge of marts with me.

Some thirty auctioneers or sons of old friends who were auctioneers helped including James Stephenson of York, Richard Turner of Bentham and Chris Cundall, all sons of men whom I held in high esteem. But all the others made great contributions – I wish I could have remembered or incorporated more of what they told me and I apologise, in advance, for my omissions and errors. My notes on Cheshire were greatly assisted by Gwyn Williams and John Wilson.

A few ladies were extremely helpful and patient and deserve to be thanked, namely Mrs Judy Middlemiss, Mrs Daphne Smithson, Mrs Margaret Porter and Pat and Katie Fieldhouse; together with Janet Allman from Beeston Castle

My very sincere thanks to everyone who has made a contribution and put up with my inquiries.

Others who make my books possible are my son, John, who assists by editing and ferreting out much rare material, not to mention stories from Longtown Mart, where he and my other son, Stuart, are the managing directors. Also Kirsty Collin, who translates my script into good sense and continues to be invaluable to me and our saleroom; my daughter, Patricia, who deals with the photographic material, and my granddaughter, Joanna Nelson, who is a dedicated researcher and a tireless chauffeur.

PROLOGUE

A number of my friends ask, "Why do you give up so much time to telling the history of auction marts?".

Many years ago, before 1950, I and a friend were entering Edinburgh's venerable and largest bookshop across the street from the Old Quad of the university. He commented that every interest was covered on the shelves, and I rashly bet him half a crown that one subject would be missing. An elderly assistant asked what he could point us to and I told him to show me a history of auctioneering, after repeating my request he went in search of the owner, who was most upset that he did not have such a rare classic.

My friend, Sam whose father was a land agent, recovered from the shock and led me to a pub across the road to buy two pints of Younger's Ale and consider what should be done and happily at the bottom of our second pint, I had agreed that there was no alternative but that, when time allowed, I would write the story of Scotland's marts. Forty years later "Ring of Memories" appeared to cover Scotland, followed by "Farmers Went to Market" to cover the English Borderland down to Kendal and now "Bid Up Now" to provide some information on Lancashire, Yorkshire and Cheshire, before, in my ninety-fifth year, I finally run out of steam.

Chapter I
A FAIR MARKET

Almost every town of any antiquity in the northern counties of England held three or four fairs each year when cattle, horses, sheep, pigs and everything required for man or beast were sold. Rules of the Fair laid down instructions including regulations as to in which street and where in that street the sellers of each commodity were to stand. At some fairs, householders would erect pens in front of their property and charge stock owners for their use, which resulted in every country town having several exciting days each year on a fixed date when fairs were held to conduct business but with the trade, fun and celebrations always incorporated into a religious ceremony to provide cash for the needs of the church and priest.

Fairs became established in the medieval period as a means to bring together the greatest possible number of buyers with the greatest possible number of sellers, a successful formula which applies to auction marts to this day. An important secondary benefit of fairs was that a transaction was made in public. A handshake before witnesses was verifiable, avoiding the need for a written contract at a time when many could neither read nor write, whilst guarding against a possible later accusation of theft or receipt of stolen goods being brought against the buyer. The temporary courts set up at the principal fairs became convenient places for settlement of commercial disputes as well as for dealing with alcohol related mischief.

At every fair a toll was charged either by the Lord of the Manor or by the Town Council. It was usually one penny[1] per beast, two pence per horse and five pence for twenty sheep, but there were, of course, variations from town to town. The authorities at fairs checked the accuracy of traders' weights and measures as these varied widely, in particular the bushel measure had many variants.

There was keen competition between fairs and the Lord of the Manor or town authorities would offer inducements such as reduced tolls, free grazing and even prizes to encourage sellers to send their livestock. For the Lord of the Manor or Town Council, fairs were an important status symbol, just as, say, the ownership of a football club

1 There were 240 pennies in £1 until 1971 when auctioneers had to learn the new language of decimalisation. No more shillings!

might be today. As important as a jewel in their crown as for the profit or commercial benefit.

The eighteenth century saw some of the smaller fairs fall away and there was a concentration at the larger ones; also some of the larger towns held a weekly market, not a sale by auction, as well as periodic larger fairs. There is a distinction between fairs and markets. The former served a regional and national market; they were large occasions and many played a significant part in the national economy. Markets, on the other hand, were local affairs which served the needs of the local population. Auction marts would, in time, cross the boundary between these and would serve both local and national demands.

Leyburn, the October Fair 1905

Convenience to a market, and a market where the buyers were reputable, was a significant factor in establishing the value of a farm. J. Anderson, writing in 1800[2] stated that farms "which are physically alike are not practically alike":-

"In one county, proper and well frequented markets or fairs for sales of livestock have long been established so that animals of every denomination may be disposed of at an adequate price for ready money, at almost any season of the year, without difficulty; whereas in the other, it is only beasts of a certain description that can be there sold, at particular times, and to a

2 J. Anderson, Essays relating to Agriculture, 1800.

few individual persons, upon credit, who have it in their power not only to keep down the prices, but also to become bankrupts, when they please, and defraud the poor sellers of the stock to a very great degree. Wherever this practice prevails, it is obvious that the tenant cannot afford to give half the rent for ground of the same quality that he could have done in the other district".

Anderson goes on to comment that there are many of the "distant hilly parts of Britain whose prosperity is much repressed by want of markets" and Arthur Young, Secretary to the Board of Agriculture had previously advised that anything over ten miles from a market, which was a day`s round trip, warranted being taken into account when the rent of a farm was being considered. In practice, he was perhaps a little cautious but the principle was correct. About thirty five miles was about the maximum distance that a horse could accomplish in a day[3].

From an early period, most parishes would have one or two small-scale dealers who often took stock to larger fairs where the "drovers" or principal livestock merchants would buy in large numbers. They then sent one of their responsible foremen or topsmen to move each drove on, often several hundred miles, to the grazing area where they hoped to sell. From the author's home county of Dumfriesshire, for instance, the majority went to East Anglia and the area north of London where they were grazed for the London meat trade.

The fairs were organized along each drove route so that the cattle sold in the Western Isles of Scotland in July moved on southwards through a succession of established fair stances, which fitted in with the great plan of the animals reaching their destination at the right time of year. As an example, the Rood Fair at Dumfries, often called the Brough Day, was held in the last week of September, when 7,000 cattle could stand on the Whitesands beside the river. Seventy miles south, Brough Hill Fair in Westmorland was eight days later and so on by stages, down country to St. Faith's Fair, 4 miles north of Norwich, which as the largest fair in England served the wonderful lands of East Anglia where the animals were fed for the London market[4]. At each fair, the animals might change hands and could have four or five different owners before ending their days at Smithfield or a similar metropolitan market. St. Faith's Fair was unique in that it continued

3 G.E. Mingay (ed.), Agrarian History of England & Wales, vol. VI.

4 In 1844, it took three weeks to clear all the livestock at St. Faith's. Its decline was swift; the last fair here being in 1872.

for a fortnight from the end of October with fresh cattle pitched each day alongside those not sold on the previous days. The last large fair of the year in East Anglia was held at Hempton Green near Fakenham in West Norfolk on 16th November and that's why the October Fair at Dumfries, which offered cattle destined for it, was traditionally called Hempton Wednesday A report of 1861 tells us:-

Dumfries Markets – October 13, 1861

Cattle – *This being the show day for Hempton Green, there was a large number of cattle on the Sands. The morning was clear and sunny, and business moved briskly. Prices on average were rather in advance of those ruling this day week. The supply was nearly twice as large as last day, and the quality of the stock was superior. The numbers were – of two and three year old Galloways, 740; of one year olds, 380; cows, 69; Ayrshires, 40; West Highlanders, 560; in all, 1,779 head. There were a good many dealers present, and a number of lots had changed hands by mid-day. A prime lot of two year old Galloways sold at £12 each.*

By the 19th century Dumfries held a weekly market or fair with large numbers of cattle converging every Wednesday, including as many as 1,000 Irish cattle driven on foot from the ships 80 miles away at Portpatrick. 20,000 or more head came annually from Ireland to Scotland in the years up to 1825 but thereafter steamships started taking them direct to Liverpool and elsewhere, where they joined the procession towards London.

Until the arrival of large numbers of Irish cattle in 1784[5] the core of the droves of cattle passing through the North of England originated at the great Falkirk Tryst. To understand the huge and important droving industry the following description of that event, although far outside the bounds of this book, may be interesting.

Henry Hall Dixon of Carlisle, "The Druid", wrote in 1860

"At the Falkirk September Tryst Mr John Murray, salesman in Edinburgh, was agent for 22 farms who sold 43 lots – a total of around 20,000 sheep. He bought Cheviot lambs at Melrose for his clients who sent them back to Falkirk as three year olds. The largest lot he sold at Falkirk was 3,679 from Cameron of Locheil's farms. The Swans, father and two sons, who became

5 Government restrictions on Irish cattle movements were removed in that year.

great auctioneers, sold maybe 10,000 sheep and 1,400 cattle in two days of the tryst."

The October Falkirk fair brought together up to 80,000 ewes and wedders (a castrated male sheep) with the principal buyers coming from Cumberland and Dumfries and Galloway.

Thomas Gisborne visited Falkirk in 1849 and wrote that the three biggest fairs in England, Barnet, St. Faith's at Norwich and Hey Hill combined were no parallel to the size and bustle and excitement of Stenhousemuir near Falkirk, where a two hundred acre field held up to 40,000 cattle with the dealers, mostly on horseback, searching out the lots they required. The Scottish sellers were mostly on small shaggy ponies that were quite at home among the cattle. On an area between the cattle and sheep were little green painted wooden sheds for each bank, where considerable sums of money were exchanged for the buyer's bills of credit. Round the outside of the field were sixty or seventy tents selling provisions and the much-consumed spirits. The owners of the tents paid a rent of two shillings and sixpence at the August fair and four shillings and sixpence at the September and October fairs. In one of the tents, the Falkirk Bank did a large business and, as most of the purchasers made payment in these tents, they were always full of cattle dealers, butchers, drovers, auctioneers, pedlars and beggars mixing together with an indescribable clamour. When a deal was made, tar dishes were then obtained and the purchaser's distinctive mark or buist (pronounced bist) put on the animals before they were driven from the field.

By 1870 salesmen had taken over the selling from the owners although such sales were still by private deals and not by auction. The following three salesmen were each to develop a large auction mart business:

Messrs Murray & Todds, Edinburgh sold 17,000 sheep

John Swan & Sons, Edinburgh sold 6,000 sheep

Mr Buist, Edinburgh sold 2,000 but also had an auction ring for the sale of rams.

The Falkirk Fair was still immense in 1850 when John Wilson started auction sales in the town but it was only after 1875, under Thomas Binnie, that a thriving weekly auction mart was established. Its particular claim to fame in the annals of auctioneering is that it became, in 1896, the first sales centre in Britain where all the fat cattle were weighed as they entered the auction ring and were then sold at a price per hundredweight. Within ten years, most English marts had followed this example of weighing the animals but many continued to sell them at a price per head, partly owing to customer resistance. P.J. Perry, in his study of "British Farming in the Great Depression" comments on the innate conservationism of many farmers who persisted in their insistence on their visual judgment in buying and selling livestock. "It was easy to demonstrate that the butcher and the dealer, daily buyers and sellers, were usually much more skilful, but the weighbridge offended the farmer's pride."

From the Falkirk Tryst, huge numbers of livestock were driven away to England, avoiding as many tolls on the route as possible.

Many of the droves travelling south into Cumbria would cross the Solway at Bowness Ford with considerable risk to man and beast – but this was only one of the many hazards which ranged from the ill health of drover or animal to robbery and, not least, fluctuation in prices. Travelling some twelve miles each day, the finding of overnight grazing for the droves became more difficult as the common lands were enclosed and cropping spread. On the brighter side, the winter-feeding of livestock made possible by the growing of turnips ensured that there were fewer deaths of lean cattle in the April droves. Some overnight stances attracted business by giving credit, which allowed the drover to pay on his return journey after selling the cattle. Drovers who called in April when the droving season commenced but before the grass had grown, were provided with hay. In the 18th century a night's grazing cost around one and a half pence per head[6].

Overnight resting places where droves could be fed, watered and kept secure were not exclusively provided by inns and village communities but many estate landowners also found it to be a useful source of income. A continuous problem, of course, was that arrivals and departures were concentrated on the nights immediately before and after large fairs.

The drovers were most important contributors to the economic success of the agricultural industry. Their numbers included landed gentry and many reputable and very astute people but then, as now, there were a few rogues who brought the profession into disrepute and caused considerable losses to the farmers who trusted them. Often only ten or twelve percent of the agreed purchase price was paid "up

6 Cattle Droving through Cumbria by Peter Roebuck, Bookcase, Carlisle, 2016, for this & other information.

front"; the rest was purchased on bills which became due to the seller only when the cattle were sold on by the drover. During the reign of Queen Elizabeth it was decided that dishonest rogue drovers must be put out of business. To achieve this, it was enacted that all cattle and sheep drovers must have a licence. The licence badge had to be worn on the upper arm at all times and be visible for all to see. Among the conditions of licence, the drover had to be a male over thirty years of age, married and a property owner, which ensured that he had some assets and an address where he could be found.

One hundred and twenty years later, Queen Anne tightened the law in England by forbidding any drover from declaring himself bankrupt without a full court hearing and, when newspapers began to print local news, after about 1800, reports of these compulsory hearings provided interesting information about the finances of drovers who admitted that bookkeeping was often not their strong suit.

A much later example of such an examination on 22nd April, 1842 in Alloa was the appearance of Bryce Johnstone, a cattle dealer from Kirkcudbright. He submitted a statement of his affairs and offered those unsecured creditors whose cattle he had bought a payment of only one shilling and sixpence per pound. In addition, he would pay all the expenses of his sequestration and the remuneration of the trustee. This abysmal offer was increased to 2 shillings (10%) per pound. The sellers would probably have only received 10 percent of the price before parting with their stock so that they would stand to lose eighty percent of the value; a crippling loss for any farmer.

Not only market forces but also cattle plague or Rinderpest could have serious implications. Such disease had struck in 1746 and the following is largely from the work of A.R.B. Haldane[7], taken from letters in Register House, Edinburgh.

Honesty and fair dealing in the face of difficulty existed long before the nineteenth century; as is shown by the letters of Thomas Bell, a drover, addressed to his partner Bryce Blair of Annan in 1746 on whom he had drawn three bills for 1,449 cattle purchased. Bell had taken 1,481 cattle south to the East Anglian markets, paying for them with the usual three months' bills. There, the cattle disease which raged that autumn throughout England attacked them, and his drove died first in scores and then in hundreds. 'God knows what

7 Haldane's "The Drove Roads of Scotland" is an invaluable reference, as is "The Drove Roads of Northumberland" by Ian Roberts. For much more on the fairs and marts of Scotland, see the 3rd edition of the present author's book "Ring of Memories".

we shall do', he writes from Norfolk on 20th December, 'We cannot get money to bear pocket expenses; all manner of sale is over...our conditions are such that several drovers have run from their beasts and left them dying in the lanes and highways and nobody to own them'. So bad was the disease that the drovers had been forced to guarantee the beasts for a certain time in the hands of the purchasers, and Bell reports that beasts already sold by him had died within the period of the guarantee. Ten days later, the position was worse. 'All is over now. We can neither pay London bills nor nothing else. We have over £1,000 charges to pay in this country and not a shilling to pay it with'. By 7th January, 1746 the disaster is complete. 'I am positive we have lost £3,000 by (the disease) already. I shall be home at Candlemas and people may do with me, as they will. They shall get every groat we have and we can do no more'.

This disease came in cattle brought from Europe to Essex and it took over ten years to eradicate with movement controls only finally being abolished in 1757. It was called Rinderpest from its German origin and death of the animal usually ensued in about ten days.

The drovers generally travelled about twelve miles per day with a break of two hours at noon. Their lads would be paid about three shillings[8] per day and one shilling for watching at night.

Traffic was also a problem and, on one occasion, when a large drove had almost crossed the Eden Bridge at Carlisle, an impatient mail coach driver galloped on with horn sounding and in the ensuing mad stampede the bridge parapet collapsed with men and animals jumping into the river. No men were injured but several cattle were lost.

Considering the large number of people involved in the droving trade comparatively few injuries were reported and it was a favoured occupation for hardy youths, known as drover lads, who enjoyed a bit of adventure. Those from the West Highlands and Islands had many a dour struggle to get their stock to swim across the many seas and lochs and rivers, which blocked their road between Skye and Stirling. Then as they left Scotland by fording the Solway to avoid tolls and extra weary miles they came across those most dangerous tidal sands where Sir Walter Scott tells us many cattle and some men were drowned in the fast-flowing tides.

Gathering of livestock at fairs had been established since the Middle

8 The shilling (abbreviation shg.) – twenty of which were in the pound, with 12d (12 pennies) to each shilling. Two shillings and sixpence is conventionally shown as 2/6d.

Ages but moving them cross-border and for long distances developed substantially only after the Union of the Crowns of Scotland and England in 1603. Before that theft of livestock either by rievers, marauders or hungry people was part of country life and access of English drovers to Scotland was inhibited not only by social conditions but the absence of a common currency. It appears that the number of cattle on the move increased approximately three-fold in the period immediately after the union.

The Carlisle records of 1662 report that 18,574 cattle passed through that city and that is without a record of those who avoided payment of the tolls.

The cross-border trade received further stimulation by reductions in Borders customs duties and their final abolition at the Union of the Parliaments in 1707.

Gisburn, the last Fair 1911

Chapter II
ENGLISH FAIRS

One of the largest and oldest of the northern fairs was held at Brough under Stainmore in Westmorland, latterly on the last day of September and the following day where, from the end of the eighteenth century until the mid-nineteenth century, there could arrive at Brough Hill up to 8,000 cattle, 1,000 horses and 15,000 sheep. The Lord of the Manor was never able to count them so he charged a toll based on so much per drove or consignment belonging to one man. It was reckoned that you could buy anything at Brough Hill Fair but cattle were, for many years, the financial mainstay of the event and the seventeen inns in Church Brough and Market Brough were quite inadequate to the demand for beds and refreshment. The event fell away after the 1850s and by 1865 the local newspaper was reporting that the livestock attendance at the fair was the smallest ever seen and that farmers were now moving their stock direct by rail and avoiding the need to stand at Brough Hill. By the mid-1870s, there would be no cattle except for a few pens where local farmers could buy beasts to fatten for the local butchers. Brough Hill remained as a significant horse fair, effectively up until the Second World War. With "more bad horses than good ones", great caution was required. The regular buyers included the agents for northern mining companies looking for pit horses from among the many Fell and Dales ponies on offer.

Stagshawbank near Corbridge, now on the A69, and close to the Roman Wall had become one of the largest fairs in England by 1794, with 5,000 cattle and, amazing to recall, over 100,000 sheep on the ground, according to the 1805 report of Bailey & Culley[9]. It had been in operation since at least the thirteenth century. Here many drovers from Scotland would add to their herds as they moved slowly south.

In 1698 a dispute arose about the ground on which the fairs had been held in times past and in 1701 the court ordered that all gates, hedges and fences interfering with the fair ground on the moor should be removed and thereafter the fairs continued unimpeded until 1824, when a number of people became dissatisfied with the conduct of the fair by the servants of the Duke of Northumberland in his capacity as

9 Bailey & Culley, Travel Journals and Letters

Lord of Ridesdale, which Lordship collected the tolls[10].

A move was suggested to a new site at Holystone, north of Rothbury, where many roads meet. However, the Duke, resenting the loss of some of his income from the fair, raised an action against the landlord there and this legal case to establish who held the right of holding a fair moved very slowly over a period of years. Dissatisfaction continued and in 1830 a few users of the fair at Stagshawbank advertised their intention to move across the road into a different parish hoping to avoid the heavy tolls, whereupon the Duke replied with the following advertisement in the Newcastle Courant on 9th October, 1830 and his steward informed all tenants and tradespeople of his Grace's opposition to this move.

"Whereas an Advertisement has lately appeared in the Newcastle upon Tyne and some other provincial newspapers, dated Hexham, September 30th, 1830 and stating that:

STAGSHAWBANK TRYST

"Will be held in the same manner as that of last year, on the 25th Day of October next, on the higher or North-West side of the Watling-street, being that Part of Stagshawbank which is in the Parish of St John Lee, and not within the Manor of Corbridge."

And whereas such Tryst, if held, will be without the Permission or Sanction of his Grace the Duke of Northumberland, Lord of the Manor of Corbridge, and Owner of two ancient Fairs accustomly held every year on Stagshawbank aforesaid.

Notice is hereby given that all persons who shall, on the said 25th Day of October, or at any time or times hereafter, assemble or meet on Stagshawbank, or any part thereof, for the Purpose of, or under the Pretence of, holding any public tryst, show, fair or market, for the sale of horses, horned cattle, sheep or pigs, or other articles whatsoever (except only at the said two ancient accustomed fairs held by his Grace on the Eve of Pentecost and Midsummer in every year) shall be proceeded against at Law in such manner as the case may require.

Needless to say the Duke prevailed and his Stagshawbank Fair continued in its accustomed way until the auction marts progressively attracted all the stock and it finally closed in 1890.

10 The Estate steward collected tolls of 4d for cattle and 8d per score of sheep at the October Fair, but the tolls at the other two fairs were only half this amount. At all fairs 2d was charged for each seller of goods or food

Skipton, a wet market day in the High Street

The fair and the marts overlapped in 1883 in the Court at Carlisle when the prosecutor said that William Bryson, a Carlisle cattle dealer, had failed to sell 24 cattle at the Autumn Fair at Stagshawbank and that he had then entered them at Mr Harrison's Mart in Carlisle, knowing full well that movement into Cumberland was prohibited owing to an outbreak of Foot and Mouth Disease. The judge took a dim view of this movement and fined Bryson £5 per head, a whopping £120 in total, which would send a clear message to other dealers – don't do it.

Along the Scotland – England border were over a dozen recognised crossing points as well as several other more secret routes. In the livestock trade, by far the busiest crossing was around Gretna where the toll bar was at Allison's Bank on the River Sark. To avoid paying a toll many droving routes by-passed the toll gate, although many toll houses encouraged conformity with the law by having a licence to sell liquor, which provided modest good cheer and welcome for the traveller.

Carlisle, the first major town encountered after crossing the western border, was an important market with the cattle standing there on the Sands, as they did until 1970. Sheep stood at Bitts Park, only a few hundred yards from the bridge and under the Castle walls. Latterly the Sands market was restricted to private sales of Irish cattle and auctions there ceased as the auctioneers found bidding was furious as drovers and dealers bid up the price but the animals were then passed out unsold by their Irish owners to be sold privately afterwards on the strength of having achieved a certain price in the ring and yet avoiding the auctioneer's commission. Similar practices have been known to pertain at many auctions of Irish cattle.

From Carlisle, there were busy routes - east by Brough, south-east by Kirkby Stephen, Hawes, Sedbergh or Kirkby Lonsdale, or southwards to Liverpool and Manchester. Those travelling eastwards or direct from the north east would head for Northallerton, on to Boroughbridge and Doncaster and then down to Grantham where the route divided, one way to East Anglia and the other taking them on south towards London or to the great fair at Market Harborough.

Skipton Fair, 1900

Robert Trow-Smith in "British Livestock Husbandry" points out that this long chain of transactions was largest in the period when urban consumption was growing rapidly but when fast animal transport had not yet arrived to bring the carcase, rather than the live animal, to London and the other great cities. In 1750, it was reckoned that 71,000 cattle had passed through London's Smithfield Market. By 1793, this had increased to 116,000 cattle together with 729,000 sheep. Of these cattle which had been driven from Norfolk into the city, it was reckoned that three-quarters were of Scots origin.

In addition to the London bound traffic, the turn of the eighteenth century into the nineteenth saw the graziers of Lancashire and Yorkshire buy many cattle from the drovers to feed the growing industrial towns of the North.

Masham Fair, 1905

Ormskirk was not only the site of a large cattle fair but also one of the earliest weekly markets to supply beef for Liverpool, only ten miles distant. A new Liverpool cattle market was erected in Lime Street at the start of the nineteenth century.

Birmingham was another significant destination and it should not be thought that all droves were "long distance". Indeed, as Peter Roebuck points out, the drift of recent scholarship has been to emphasise the importance of intermediate fairs and markets in the north and midlands where, even in the 17th century, good profits could be made. Many droves would be of medium or short distances and I recall my grandfather telling me that the Scots drovers would often only travel as far into England as Brough Hill before turning for home. It is probable that the Brough autumn fair was their furthest point of travel from which they would be able to make it home before the winter weather made the northern Scottish roads impassable. Some drovers to the later season East Anglian fairs would stay the winter in the better climate and return north in the longer days of spring.

Arthur Young wrote extensively on agriculture. In 1770 he visited

Yorkshire and left some information about the drove roads and fairs. Later newspapers reported that Malton was among the largest of the autumn fairs with very large numbers of sheep. Yarm was another where the third day was dominated by 700 carts of cheese. By 1822, the Newton Fair near Manchester caused chaos on all the roads and strong Galloway cattle made up to £9, about 10 shillings less than the previous year. Askrigg, east of Hawes, was a popular fair from early times with a large area providing cheap grazing for cattle and flocks of sheep, which mainly came from Swaledale to be sold.

Middleham Moor, near Leyburn, reported in 1870 a pretty good show of Scotch and half-bred sheep with great quantities changing hands. On the second day there were "many noble droves of cattle and the best were soon disposed of at good prices". The third day was mostly horses with Shetland Ponies making up to £10.

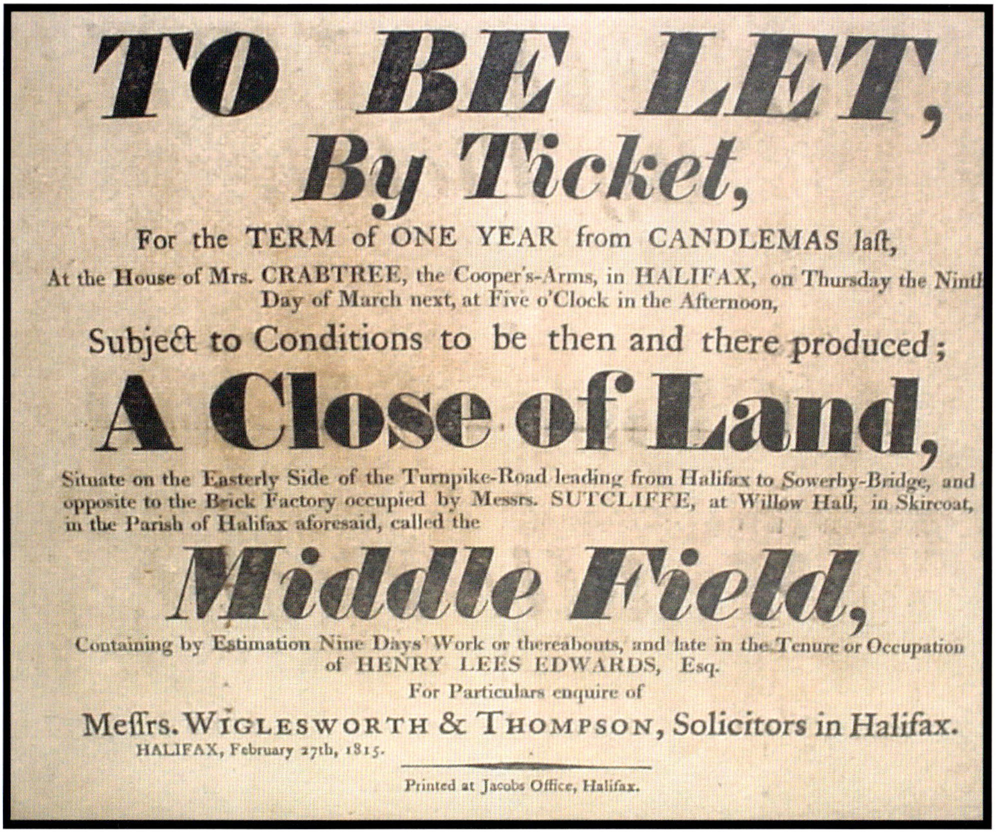

TO BE LET,

By Ticket,

For the TERM of ONE YEAR from CANDLEMAS laſt,

At the House of Mrs. CRABTREE, the Cooper's-Arms, in HALIFAX, on Thursday the Ninth Day of March next, at Five o'Clock in the Afternoon,

Subject to Conditions to be then and there produced;

A Close of Land,

Situate on the Easterly Side of the Turnpike-Road leading from Halifax to Sowerby-Bridge, and opposite to the Brick Factory occupied by Messrs. SUTCLIFFE, at Willow Hall, in Skircoat, in the Parish of Halifax aforesaid, called the

Middle Field,

Containing by Estimation Nine Days' Work or thereabouts, and late in the Tenure or Occupation of HENRY LEES EDWARDS, Esq.

For Particulars enquire of

Meſſrs. WIGLESWORTH & THOMPSON, Solicitors in Halifax.

HALIFAX, February 27th, 1815.

Printed at Jacobs Office, Halifax.

The Municipal Corporation Act of 1835 gave much needed wide-ranging powers to local authorities to control streets and public places. It was used for the provision of street lighting, controlling the selling activities in the streets, paving streets and lots of other would-be improvements. Bridges and the widening of streets were also covered by the act and it had a considerable impact on the locations of fairs and markets. Slaughter houses and butchers` shambles were generally moved to a more private area where cleansing and draining were improved.

By 1870, the Halifax Great October Saturday Fair had been moved from the streets of the town to a market area so that the confusion and congestion were avoided. Cattle and horses were numerous and pigs were forward in record numbers with high prices ruling.

Beverley November Fair that year lasted from Tuesday to Saturday with very large numbers of horses shown and many bought for the army cavalry. Cattle prices were also reported to be higher.

Doncaster October Fair in 1870 was cheered by the knowledge that the supply of winter keep was now assured and this breathed new life into the trade with the following prices reported:-

Milk cattle £16- to £20-; heifers £12- to £16-; steers £9- to £10-.

Sheep 35 shillings to 45 shgs; lambs 20 shgs to 28 shgs.

Small pigs 20 shgs; store pigs 35shgs to 45shgs; pork pigs 80shgs to 100shgs each.

Otley, The Licks Cattle Market, 1884

The power to charge tolls, granted by ancient Royal Charters, continued to penalise some auction marts, even when sales were held in private premises, until well into the twentieth century but by around 1930, the auctioneers were challenging the Lord of the Manor or the Town Council and contending that, as the mart was no longer held on the streets of the town but in a privately-owned area, tolls were not applicable. For example, in Otley a charter of 1222 continued to extract money from the auctioneers until 1935, when Mr Lister challenged it. Tolls, of course, continued to be levied where the local authority owned the market.

When the widespread enclosure of land meant that no more free or cheap grazing was available for droves, the drovers were encouraged to rent fields for the seven-month grazing season, which was much easier for the landlord who would often issue tickets before the letting auction to approved bidders only.

In 1815 the French were defeated at the Battle of Waterloo followed by the immediate demobilising of many soldiers and sailors who no longer needed huge quantities of meat from government stores. Between buying in July to selling in October, which was the normal practice of dealers, prices of livestock were halved and half the drovers of cattle were bankrupt. They had really met their "Waterloo"! As the century progressed, the auctioneers would slowly take their place with the age of transport by steam giving their sales a great impetus so that by the 1860s auctions of stock from multiple sellers grouped at a mart were beginning to be widely recognised as the best method of sale.

Single seller sales by auction of livestock or chattels on the client`s property had, of course, been common for a considerable period and in the late 18th century the terms might be set out as in this advertisement from 1790:-

"Ready money will be expected for every article under forty shillings value and for all above that credit will be given to Candlemas next, on the purchaser finding security".

The railways affected the fairs, which for hundreds of years had been the only way for stock breeders to sell their stock and realise its value in dealing with traditional buyers. Now the steam train could bring in unknown dealers and just as quickly whisk the stock away to unknown destinations. Many stewards of landed peers on whose land fairs were held were also surprised to read in the new local newspapers a week before the fair the discouraging and often inaccurate information that "stock prices were falling and many fairs had recently sent many lots home unsold". To counteract this fake news, the House of Lords heard many appeals from peers for full and accurate reports to be published as an essential guide for all involved and requested their Lordships on whose estates fairs were held to send an under steward to collect and publish full reports of the trade

Colne, going to the market
(Edward Walton Collection)

These published reports brought about a marked improvement in necessary knowledge, for both sellers and buyers, which continued until the auctioneer took over and farmers at last knew if prices were up or down. However some farmers suspected, and perhaps still do, that auctioneers' reports sometimes gave an optimistic view of prices and an exaggeration of the numbers sold. In the twenty first century, the scope for such exaggeration is greatly reduced owing to computer generated reporting but the auctioneer, by his very nature, remains an optimist.

Chapter III
DROVING DAYS

A Yorkshire cattle dealer, John Birtwhistle from Skipton, North Yorkshire was held in such high regard that the British Linen Bank (one of the largest in Scotland), when instructing their cashier before he went to the Falkirk Cattle Tryst, laid down that the only person that they were to supply with cash was John Birtwhistle – up to £2,000, if he required it - otherwise the cashier was only to negotiate bills for drovers who were known to the bank.

When, in around 1740, John Birtwhistle attended the great Falkirk Cattle Tryst or Fair he met Isobel Shearer in Falkirk. They married in 1741 and she bore him three sons, who all became successful cattle drovers with records of buying in many of the estates and islands of the North West of Scotland. They even traded as far out as Harris, where they paid all their purchases to the laird's factor or agent, who then deducted the rent before paying the crofters and small farmers for their cattle. The factor did not allow any other cattle buyer to visit the island, nonetheless Birtwhistle was reputed to give decent prices, which he disclosed to each seller so that they knew how much to expect to receive in due course from the factor after deduction of their rent.

In the fifty years or so up until about 1810, John Birtwhistle and Sons were amongst the largest of the north of England drovers[11]. John Birtwhistle's home farm, was at Long Preston in the Yorkshire Dales but he bought farms at Gatehouse in Galloway, at Dalbeattie in Kirkcudbrightshire and in Lincolnshire; presumably staging posts as he moved cattle from Ireland, through South-West Scotland to Yorkshire and on to Norfolk and Suffolk. His sons held five sheep farms in the Fort William area of the West of Scotland, presumably acquired when they went north each June to buy the Highland kyloes. Birtwhistle rented 700 acres from a Mr Lister at Malham[12], near Long Preston and it was claimed there could be several thousand cattle held there at one time with some 20,000 passing through over the summer. Three Birtwhistle sons were taken into partnership to carry

---

11 Tony Stephens, Dumfries & Galloway Natural History and Antiquarian Society, Transactions, 2009.

12 A visitor to Malham Fair around 1830 found every corner full of cattle and every garden a pen for sheep. Worse still, no accommodation was to be had, not even a chair near the fire. He soon moved on before a large dance started in the evening.

on this successful business, which was one of the first to advertise their own sales in the newspapers and periodicals and, thus, not be totally dependent on selling at the regular fairs, as most drovers were. In March 1782, they advertised that, on the 14th and 15th of the month, a large drove would arrive at Hoxne in Suffolk and would stand there until sold. In December of the same year it was announced that a drove of Irish and Scots cattle, "the property of Messrs Birtwhistle" would arrive at Handford Hall, also in Suffolk, to be sold. William, the last of John's sons died in 1819 and the Leeds Mercury reported that "the greatest dealers and graziers in the kingdom are all extinct".

The Birtwhistles had found the little village of Gatehouse in South West Scotland, midway between the points of landing for Irish cattle and the market town of Dumfries which met the great drove roads from the North, to be an ideal headquarters for their cattle trade. They contributed greatly to the prosperity of Gatehouse and brought experts from Yorkshire to set up two mills in their village, one for wool and, by 1788, another for cotton. These provided work and, with support from the local landowners, the village soon doubled in size.

Now Gatehouse is a popular holiday destination where visitors can walk down Birtwhistle Street, where this Yorkshire family built a row of houses, each with one large space upstairs where spinning took place. Their Manager's house still stands on the corner.

Many crofters worked in the mills there leaving land vacant for large cattle parks whose owners sold their stock to the Birtwhistles. Alexander Birtwhistle had a large house in the High Street, later converted to the Bank of Fleet Hotel.

Then as now, the newspapers and magazines carried market reports. The Farmer's Journal of 30th April, 1827 reported that the cattle trade in Cumberland had collapsed. At Carlisle Fair on the Saturday, large numbers had been forward and buyers were numerous with numerous sales early in the day at high prices. At Penrith five days later it was a different story with cattle prices down fifteen per cent and not half the stock were sold. Sheep were numerous and few sold; but a farm dispersal by auction enjoyed good prices up to £24 for fat bullocks.

Morpeth was an early example of a weekly market where a large number of stock were walked each week from Berwickshire and Roxburghshire taking several days on the journey to Morpeth. The Berwickshire cattle would leave on the Friday to catch the Wednesday market. Badly driven cattle would lose at least half a hundredweight

(56 pounds weight, about 25 kilos) on the journey and, if sweated, would lose that "bloom" on their coats which so helped their sale. The best cared for were often those where at least one of the drovers was a member of their own farm's staff. Many animals came from further afield, including Cumberland. 200 cattle could stand in Morpeth town square and sheep were penned in the streets. Sixpence was charged for the use of a pen. By 1800, about 100 cattle and 2,000 sheep were sold each week making it one of the largest weekly markets in England.

When John Church, a leading Annandale farmer, had a few score of two shear sheep ready in May, June or July he always insisted on droving them the sixty miles to Morpeth Market, where, for the Newcastle butchers, mutton was never considered too fat to meet the appetites of the pitmen and shipbuilders and there were always plenty of buyers. He always sold them by the "clad score" adding one for luck to each score or batch of twenty sheep.

The state of the public roads was a constant problem to the drovers as much as to the travelling public. They were initially maintained by each parish using statutory labour. This meant unwilling, unpaid and unskilled workers under often unskilled supervision and often only working for some six days each year. The change to the toll-charging system introduced in around 1750 was far from popular but it did bring about many improvements with Turnpike Trusts maintaining, improving and charging for the roads. The first Acts of Parliament to construct turnpike roads in Yorkshire was in 1750 and road tolls, vexatious to the drover, continued for about a hundred and thirty years. The toll in 1752 was ten pence per score (20) of cattle and five pence per score of sheep. In 1734, the coach journey from London to Carlisle took nine days. By 1773, this had been reduced to three days. Better vehicles and better management of the horses, as well as the improved road surface, all contributed to this substantial improvement. Toll gates could be frequent on any journey and, for instance, between Penrith and Burton in Kendal there were five toll gates in only 32 miles. Steam road rollers were not introduced until as late as 1886. They provided a great improvement to the road surface but this came too late for the droves where lameness had been a great problem. Many blacksmiths had specialised in shoeing cattle. Two small plates, known as caulkers in Scotland, were put on each front foot. Back or hind feet were seldom shod and the shoes could last for three or four months.

In 1810, Thomas Telford, possibly Britain's greatest engineer, was employed by the Government to improve the Glasgow to Carlisle road. He built bridges over the Sark and the Esk, thus reducing the distance from the Border to Carlisle by five miles. There were no longer detours by Longtown and the new toll house was built at Sark Bridge, where it still stands, famous as the last house in Scotland and now a coffee house. Two thousand cattle a week passed Gretna in the autumn with the Eden Bridge at Carlisle being the next barrier where the toll for Cattle was eight pence a head. Around Penrith collectors reported an extra thirty per cent collected in tolls on market days. By the time that marts were being developed it was widely accepted that road tolls were out of date and that the Turnpike Trusts should be dissolved. However, most of these had borrowed money to fund improvements, and with railway competition taking their business, they had fallen further into debt. In around 1850 the Government prohibited further borrowing by these bodies and road maintenance costs fell more heavily on Parish Councils. In 1870 the dissolution of Turnpike Trusts in the North of England commenced and was completed in 1883.

The general improvement of roads was, of course, a great help to the local marts and this work fell mainly on the County Councils who did a good job until the increasing numbers of motor vehicles churned up the loosely surfaced roads. In 1909 the central government formed a Roads Board to provide national support and this developed, in 1919, into the Ministry of Transport.

Drove roads, often ran separate from and parallel to the toll roads as. For example, the Bullock Road carried the droves south, roughly parallel to the A1, from Newark to Alconbury, near to Huntingdon, a distance of about sixty miles; from there, the Ely Road went east into Norfolk. When sharing a route with the travelling public, the disruption by the droves held up travellers and caused much ill-will, whether they were trying to overtake or were meeting them head on. Meeting other farmers could cause the greatest problems as stock could easily be mixed.

The following 1810 advertisement is noted:-

"On Thursday last after the Falkirk Tryst a heifer joined a drove which belong to John Creighton of Roberthill, Lockerbie. Any person proving her to be his may have her on paying the usual expenses"

There were many other such advertisements which were, of course, often never seen by the aggrieved party who probably lived many miles away.

Droving usually brings to mind cattle, but extremely large numbers of sheep were also driven long distances. McCulloch's Statistical Account of the British Empire records, in 1837, that Cumberland farmers were buying Scottish Highland sheep in large numbers and fattening them on turnips for sale in Liverpool and elsewhere. Drovers must have brought them from Scotland into Cumberland and other drovers would move them South when they were fat until, in 1847, the railway reached Carlisle and fat stock took the train.

Movement of livestock by droving had increased steadily throughout the eighteenth century reaching its zenith in about 1835 when the railways and steamships began to make long distance droving redundant. The passage by steamship from Aberdeen to Newcastle took only eleven hours. Livestock from an Aberdeenshire farm could take less than twenty-four hours to reach Newcastle Market instead of their previous twenty-four days on the road.

The second public railway in Britain was built in 1830 from Liverpool to Manchester and was such a success with the cattle trade that by 1840 over 100,000 cattle were carried from Liverpool to Manchester in one year. Many of these were Irish or American stock arriving by sea and immediately sent on by rail. Ten years later railways had reached most cities and there were some 6,000 miles of track in England and Wales; the days of the traditional drovers were numbered.

The increasing population of the 18th century had been fed by rising numbers of sheep and cattle which had to change hands more frequently. The existing system at fairs of bargaining, often by buyer and seller shouting at each other, and thus reaching or pretending to reach agreement was clearly inadequate. Leading farmers suggested that the sellers should not fix a price but should accept bids on an upward basis until a satisfactory final price had been reached and, in this way, the great bustle and excitement, which dealers tried to generate, was done away with and much more business could be done in less time and with less trouble. Selling under this bidding process meant there was less room for those disagreements, which, at a fair, always had the potential to turn into fisticuffs or even a riot. Developments were now all pointing towards the need for a livestock salesman or an auctioneer to oversee these transactions and this was soon to come.

Eric Loadman, who farmed at East Barden Dykes near Leyburn recalled how, at a 1913 fair, the seller of sheep from Coverhead Farm found late in the evening he could not find a buyer for his 100 ewes. In a Public House in the town he expressed his disappointment and a Mr Bert Blenkinsop took a £5 note from his top pocket and bought the sheep at a paltry one shilling each. One assumes that he took his sheep to the mart next year. One further example of this one-sided method of sale is told by Ken Hird from Swaledale, whose grandfather, the late George Hird, had, at the end of the day, to send his son in pursuit of the buyer to call him back and accept his offer as this one option was all he had. Little wonder then that auction marts, with competitive bidding and guaranteed payment, were welcomed.

> **NOTICE.**
>
> STAMFORDHAM AUTUMN TRYST will be held on THURSDAY, October 5th, 1871, for Cattle, Sheep. Lambs, and Pigs.
>
> Mainsbank, September 20th, 1871. [443
>
> **BLANCHLAND AUTUMN FAIR.**
>
> THE above FAIR, for the SALE of SHEEP and LAMBS, will be held on WEDNESDAY, October 4, 1871, in the CLOSE PASTURE, by permission of the Trustees of the late Lord Crewe.
>
> College, Durham, Sept. 21, 1871. [444
>
> **NOTICE.**
>
> A FAIR for the SALE of CATTLE, HORSES, SHEEP, &c., will be held at STANHOPE, on FRIDAY, October 13th next, in a field occupied by Mr R. Mews, near the Railway Station.
>
> N.B.—The Autumn Fair will be held annually on the second Friday in October, and the Spring Fair the Thursday before Easter.
>
> Stanhope, September 12th, 1871. [446

Taking your stock to the fair was the highlight of the farming year with the farmers being about equally divided between those who thought it exciting and great fun and those, the quieter sorts, who found it to be a hugely stressful event. Their first worry was when to leave home and what supplies to take as none could foretell the unexpected hold ups ranging from swollen rivers to runaway or injured animals. Nearer to the site every road was gridlocked with herds and flocks, gigs and pedlars, men, women and children, "all a-going to the fair".

At the entrance everything slowed down as the men in authority counted numbers, collected tolls and directed everything to its appointed area, which was often difficult to reach through all the stock already on the ground.

It was essential to keep roadways clear for livestock to move in or out and, most importantly, for buyers and sellers to move quickly from one lot to another or go quickly to the bank tents to settle the deal or to the bar tents to celebrate it.

When the stressed seller found a spare area there were, of course, no fences or gates to control their stock and we all know that cattle like to rush to see other cattle. To add to the pain, a seller's stock might be placed beside a really good lot which made the stock which had looked so good at home a little embarrassing. Then the question of when to sell or not to sell added to the agony of going to the fair. No wonder that there was a welcome for the auctioneer who provided a local, convenient mart where there were pens with gates to keep the animals secure from straying or theft and a system to get clear bids and a quick sale with one hundred percent of their money safe. And, for good measure, your stock ably described such as "from a hard Pennine farm, these will grow like mushrooms for you"! The auctioneer`s patter is the one thing that hasn't changed with the times.

The Napoleonic Wars of the early 19th century brought a good deal of prosperity to northern farmers, which enabled many to improve their land and diversify and become merchants. A few became successful dealers in livestock and many of those were to go on to be founders of livestock markets. However, as previously mentioned, the conclusion of the wars with the French was to bring financial disaster to the drovers.

Droving had encountered many ups and downs but by my own family tradition, my great great grandfather, who was then a drover, recalled that 1815 was the worst year ever experienced. Napoleon had escaped from his island prison and raised a French army which

Britain responded to by calling up large numbers of sailors and soldiers. The Government rushed to fill their stores with all the beef they could find. Prices went up so the prices of store cattle also went through the roof. After the British victory at Waterloo, at the end of June, it took only two months for the military command to decide to demobilise the troops, which left large surplus stocks of beef to depress the market. The drovers continued to buy high in July and August. September prices were still good, but by October when the droves reached the south prices had halved. Many drovers went bankrupt however my ancestor, John Thomson, assured all his customers, who were all local friends, that he would pay them in full the next year. In fact, it took him three years to pay everyone.

Most dealers or drovers paid only a small proportion of the price in cash, the remainder was covered by a promissory note or a bill payable in three or four months, by which time, if all went well, the drover expected to have sold the cattle or sheep further south. This huge business was based on trust in a predominantly untrustworthy body of men, who had to cope with so many unknowns between purchase and sale that affected the outcome of the transaction.

Between 1820 and 1840 some bad hard times were experienced in agriculture and there are quite a number of instances of the farming community coming together to support a respected farmer and his family who were in financial distress. Advertisements are to be found requesting the creditors of such a farmer to attend at the local inn at a specified time in order to concert, or agree, "upon a plan respecting his farm for he gives notice that without assistance or some helping friend he will not be able to hold the lease of his farm any longer". Usually assistance or relief was forthcoming and the farmer would survive to plough for another season.

A hundred years after this, in the hungry 1930`s, the author`s father, in common with many other auctioneers would have to be the "helping friend" to whom the hard-up farmer would turn and to whom the mart would provide sheep and cattle, together with the support and advice that kept many tenants in their farms through the hard times, enabling their sons and grandsons to remain as successful farmers today.

To return to 1840, the auction sales of Border Leicester and Cheviot rams started in that year at Kelso and by 1865 practically every fair of any repute had a corner where an auctioneer would hold an auction sale where rams were either sold or leased for a four-month season, although, by this time, the habit of leasing was dying out. Most of the hirers came back year after year and got preference before newcomers were allowed to bid.

Ten years later fairs were reported to be much reduced, owing to the popularity of the auction marts, and 2,000 rams were sold in four auction rings at Kelso.

From the mid nineteenth century sales by auction became increasingly popular with many annual sales of store sheep held in August with around two to three thousand from their district at each sale and the auctioneers offering terms such as either six months credit or two and half per cent discount for cash. At this time, a multiplicity of auctioneers commenced annual village sales of this nature throughout Yorkshire and Lancashire.

As auction marts were first launched there were no pre-ordained systems, either of business or buildings. They were open to all the influences of the universe. Yet at approximately the same moment all focused on a single aim, which was to do the job properly no matter how inadequate their premises might be. Often the disused stable yard of a once prosperous coaching inn was initially considered to be ideal. Such cramped conditions would then be cursed by the auctioneer and his staff for ever and a day and yet many such marts would continue for a hundred years or more.

Auction marts were good for democracy as lords, factors, owner occupiers, yeomen, tenants and labourers all leaned on the same pens, stood in the same ring and bid against each other and each man's stock was sold with the same attention.

Important annual auction mart sales were described as Annual Fair Day sales thus carrying on the tradition that had attracted large numbers to the event. Many of these dates had marked religious Saints days established by the Catholic church many centuries earlier and transferred to secular authorities after 1539, and many such titles still exist.

For a time, as cattle and sheep transferred to the auction system, horses remained as the mainstay of the traditional event. Among others, Beverley Fair lasted all week with horses changing hands every day. Now only Appleby remains as an annual horse and pony fair.

The declining fairs and the ever-growing auction marts co-existed until the 1880s. After that fairs pretty well ceased to be of importance to the livestock trade, but hung on as centres of frivolity and the fair day often included an auction sale with many of the largest market days still being advertised as Annual Fair Day sales.

Despite all, droving continued one way or another for a further half century or so but by the 1930s the last of the drovers were on their last legs; few in numbers but full of spirits. Their droves were usually for no more than two or three miles.

In Dumfries, latterly, a drover's longest days were mart sale days when I recall he might walk a dozen cattle in from a dealer's farm and then help at the mart until Willie, a Skipton cow dealer, arrived who would need an escort to the pub on the corner; after that he would drove cows from the byre to the ring and watch Willie's purchases to make sure they were properly milked out. He would later walk these purchases to the station a mile away. Before that he looked out for the arrival of a local butcher who needed company for a refreshment. After the sale he would bed and feed the butcher's purchases with the auctioneer's best hay and deliver them to the slaughterhouse a mile away when they were required. He would call at the auctioneer's office most days to see if there were any telegrams for him as dealers used him to meet trains on which they had dispatched sheep or cattle from distant markets and he had a wonderful knowledge of when they would arrive. He was allowed to keep warm in the railway signal box until the train arrived whereupon he would go to the loading pens.

Chapter IV
AUCTIONEERS

Sales by auction are recorded in Roman times and it is likely that some tradition of these continued over the centuries including, quite possibly, agriculturally related sales. However, until the proliferation of newspapers with their many advertisements, there is no real record for us to research but there is little doubt but that the eighteenth century saw an ever-increasing trend towards selling by auction.

By 1777, there were sufficient livestock sales held for the government of the day to consider it worth their while to tax them and a duty was introduced to be paid on auction sales of livestock which were not held on the farm of origin. Whether this was actually to raise money or was intended to discourage auctions as being to the disadvantage of the great landowners, who controlled the fairs and controlled public policy, is a moot point. Possibly it may have been perceived that auctions gave dubious persons the easy opportunity of disposing of stolen stock. All three notions have been suggested. There is, however, little doubt that, almost seventy years later in 1846, it was the desire of the railway companies to encourage the collection of livestock near their stations that prevailed and the tax of five per cent then levied on all such sales of stock was abolished by Sir Robert Peel's government.

Before the abolition of tax on sales, it is apparent that the popularity and frequency of the auctions had increased considerably. In 1778, the first year of the duty, sales of half a million pounds were recorded; by 1832 this had increased to some two and a half million pounds. At that stage the scope of the tax was considerably reduced with it ceasing altogether, as stated, in 1846. At that time, a licence fee was introduced for auctioneers.

The 1851 census[13] records about 3,500 auctioneers in England and Wales. Only relatively few of these auctioneers would be specialist agricultural auctioneers. Most would sell furniture and chattels but the eighteenth century had shown a great increase in sales by auction of property, including a number of large estates.

---

13 See the voluminous "The Agrarian History of England & Wales", volume VI, 1989 (General editor – Joan Thirsk) which has assisted with this chapter. The 1851 census also details 5,540 people, including 24 women, who describe their occupation as cattle and sheep dealers and salesmen. Also 1,016 pig dealers and a multitude of curers.

Of all the professions, that of the livestock auctioneer must be one of the most difficult. Other professionals represent one interest only, while the livestock auctioneer represents two clients who are at war during every minute of his working day.

John Thornton of London, selling the cattle of the aristocracy by auction, gave the profession an aura of respectability and integrity, which cannot be overvalued. He and the successors to his business held many herd sales on the home farms of the sellers throughout the northern counties, combined with regular collective sales at marts such as York and Crewe.

Both sellers and buyers and auctioneers profess to want a fair price, the seller a fairly high price, the buyer a fairly low price and the auctioneer a fairly acceptable price, with this irrevocable price to be agreed on average forty times an hour and calling for the firm diplomacy that good auctioneers must possess. My father claimed that great auctioneers were not made but were hewn out of the hardest rock. Thus auctioneering could be quite a hard and stressful life with many long chilling days. It is no surprise that many auctioneers died when barely in sight of old age.

When researching early 19th century sale advertisements, I have often wondered why, when farmers moved from their farms and their dispersal sales were publicised, the name of the auctioneer was never advertised. I have recently found out that, whilst every town had a cheerful auctioneer who would sell the pots and pans and feather beds and sundry tools and effects, which he sold in the most leisurely way with much banter and many lengthy descriptions, he was not yet entrusted with the livestock and important implements. These were either sold privately, on the sale day, by the farmer or by a friendly dealer. The money was then collected by a banker or solicitor's clerk, some of whom, as time went on, could see the opportunity to earn some money by doing the auctioneer's job. After about 1850, the trend was towards putting all the stock, crop and implements under the hammer. This provided competition for everything and a much more efficient collection of the proceeds; all was thus handled by an auctioneer with some professional or business background which encouraged trust and whose name was now proudly advertised. In many cases, having got a start in business, he went on to found a local auction mart.

The founders of marts had a wide range of backgrounds. Usually they had a father or grandfather in farming. Very few had fathers who had been auctioneers, possibly because traditional local auctioneers came from a lower stratum of society and that it was not an occupation that generated much profit. You needed a little cash to start a mart, as pens cost money, and you had to encourage confidence and appear to have the ability to pay the sellers, something that sometimes tested a few of the early entrants to the auctioneering business. So what occupations did the fathers of the founders of some of the country`s major auction companies have:-

Mr Atkinson was a farmer
Mr Addison was a policeman
Mr Corson, a crofter
Mr Fraser, a land agent
Mr Harrison, a farmer
Mr Hope, a farmer
Mr Iveson, intended for the church
Mr Kirkpatrick, a Scotch draper
Mr Kidd, an accountant
Mr Laurie, a banker
Mr Murray, a farmer
Mr Oliver, a solicitor
Mr Rae, a cattle dealer
Mr Swan, a cattle salesman
Mr Thomson, a farm manager
Mr Walton, a farmer
Mr Wallet, a draper

Three others, only, were sons of country auctioneers, while some had owned inns and therefore had the yard that could be converted into a mart at little expense, to replace the coaching business which had been supplanted by the railways.

The livestock auctioneers, as they developed in the nineteenth century were men with the gift of the gab and a reputation for honesty, whilst not always being great book-keepers. As can be seen they tended to have a minor professional or small business background which set them apart from the long-established small-town auctioneers of chattels, furnishings and the like.

Larger sales were usually attended by either a banker, a solicitor or an accountant to record the sale, collect the money and enable the auctioneer to report a total to the vendor or their anxious family at the end of the day. In due course, some of these professional gentlemen would organise the setting up of limited companies to operate a mart, with local farmers and business people as shareholders and themselves as company secretary.

By the 1840s a number of these auctioneers were widening the scope of their businesses, perhaps setting up a few pens at village fairs to sell the sheep off the surrounding fells or promoting specialist sales of rams.

Occasional auction sales at the historic fairs were still going on as late as 1886 when Mr William Cook of Hexham and Mr Thomas Blandford of Corbridge both advertised their annual auctions of rams at Stagshawbank on the Monday of the fair. Mr Cook, selling 120 Border Leicesters, adjoining Mrs Snowball's tent (well known for her quality of catering) while Mr Blandford offered 50 rams "at the usual place" at 11am.

After the arrival of the railways, dealers and drovers were joined at the fairs by an increasingly important species, the salesman, who acted for many sellers and buyers, albeit still by private bargain, and brought more stability to the market. Many fairs did little business until a leading salesman arrived to set the tone of the market and decide whether prices would be up or down on the previous such event. These men later saw the trend towards sales by auction and went on to establish some of the most successful auction marts.

My late friend, Jack Collard, from Wincanton told me that in looking at some of his family papers he found an interesting change in nomenclature in the decade 1840 to 1850. The title of "Drover" was dropped by the men who bought the stock and they then called themselves "Dealers", then in the next decade a number changed their job title to "Salesman". Many auction firms retained the title of "Salesmen" on their billheads until well into the 20th century.

From 1860 onwards, the title of "Drover" applied only to menial workers and those who actually drove the stock around the marts. By 1930, the Government Employment Exchanges called them cattle wallopers, which I always thought was a little unfair to a body of men who were mostly kind-hearted to their stock.

From this assortment of discordant origins, the cohesive and respected profession of livestock auctioneer was established, so that by the 1880s they were a hugely important and trusted part of the livestock farming industry that had swept away the tradition of fairs and bartering.

The old order had certainly changed. No longer was marketing a problem. If a farmer walking round his fields decided to sell one beast or fifty, tomorrow or next week, he could drive them a short distance to his local mart where there would always be a buyer or two for them and an auctioneer to sell them at a fair price and able to pay the farmer in full.

Whilst writing this, I have come to realise that there have been significant regional differences in the structure of mart organisations. In Scotland and the extreme North of England, the auctioneer was personally in command, but coming South, particularly into Yorkshire, the country solicitor or leading local entrepreneur was much more likely to have been in control when the marts were established.

In Scotland and Carlisle in 1880, the advertisements had the name of the auctioneer at the top, whereas further south the place name of the company was at the top (e.g. Bedale, Ripon, Richmond, Sedbergh, Leyburn etc). and the mart secretary's name was at the bottom.

On visits to small market towns and looking for information, I have sometimes been directed to a deceased solicitor's family or a long-established legal firm to find out about the mart. The logic in this is that marts had often been set up as a limited company, often under-capitalised, with a very busy solicitor as secretary and controller of the business and, generally supported by an extensive board of directors made up of local farmer shareholders. They would employ an auctioneer by the day who, possibly, sold in several different towns for perhaps one guinea per day and, on his free days, sold on his own account at farm sales. In Scotland and the far north of England, auctioneers tended only to work for one firm.

It is my conclusion that this structure was bad for the mart for the following reasons:

1. The name localised the business;
2. The auctioneer was usually underpaid;
3. Therefore he was keen to take on other work;
4. The initiative of the auctioneer was often curbed by a cautious director and the auctioneer was often encouraged to sell at other marts. Some auctioneers sold for three different and competing marts each week with some conflicting interests;
5. If the secretary was a solicitor then clients of other local solicitors might be discouraged from using the local mart;
6. Maximum dividends would be paid, whereas in Scotland reserves were built up and the business expanded;

7. At the first suggestion of a problem, the secretary might well distance himself, closing down the mart and selling the assets. Often getting a good price per share, but much less than the underlying value of the property for development.
8. Whilst many farmer directors did, and still do, contribute very substantially to the success of their businesses, there was always the risk of an element of self-interest in some of their decisions.

CHAPTER V
CITY MARTS

The Industrial Revolution increased the population of cities such as Liverpool and Manchester five-fold in the century after the 1811 census. Inevitably it was demand at the city markets that controlled prices in the many provincial country town markets whose stock was largely destined for these great centres of consumption.

About one hundred and fifty marts, spread across Lancashire and Yorkshire, are listed in the gazetteer section of this book. Their pattern was partly shaped by market towns, where the mart attracted trade and served the local community but the dominating factor was the huge industrial cities where large markets occupied by a wide assortment of auctioneers served a vital purpose but were often considered to be a nuisance by residents and their local council. The history of these city marts is of their gradually being relocated from the open streets and common land to a compact market site comprising pens and sale rings; often with railway sidings, a meat market and slaughter house adjoining.

Metropolitan Cattle Market, Caledonian Road, London

London was, of course, the country`s largest centre and, by 1840, the railway companies were demanding that markets there should be convenient for railway traffic. The city council decided that the live market must leave Smithfield with its narrow lanes and be re-located to Caledonian Road, Islington, where Prince Albert opened the thirty-five acre modern market in 1855. It was provided with pens for 6,000 cattle, 35,000 sheep (in 1,800 pens) and 9,000 pigs with lairages and abattoirs on the site. Railway sidings ensured that stock bought in the marts of Lancashire and Yorkshire could be in the Metropolitan Market on the next day.

Liverpool

A cattle market was held, reputedly on a very small scale, in Church Lane, Liverpool early in the 18th century. By later in the century things had greatly improved with a new market being established at St James` Place, near St James' Church. It was held in a farmyard there and the farmhouse was converted to a tavern. It must have met with general approval as a Liverpool newspaper of 1780 reports:

"It is with pleasure we can inform our friends and the public, that on Wednesday last, at the fair in Harrington, adjoining the south boundaries of this town, there was a most pleasing and numerous show of fat oxen and other cattle, together with an incredible number of sheep and lambs. There was the highest approbation of its establishment, and particularly of the accommodations both for the gentleman dealers, their customers, and their goods. The quantity of land in a good ring fence, well-watered, and necessary building for the reception of cattle, added to the vicinity of the flourishing town of Liverpool, we should not wonder if it became in a short time the first fair in England of its kind.

Cattle, sheep and pigs were sold at St James` until pens were erected at Kirkdale, near the Liver Tavern and the market was moved there. It is said that the reason for the change of location was that the great majority of cattle were, at that time, driven from Scotland and the North of England; Kirkdale, on the north road, was thus a much more convenient location. Kirkdale remained open until the move to Old Swan in 1832.

In 1820 it was, again, reported that no town in England could boast better market accommodation than Liverpool with the new cattle market in Lime Street and the pig market in Chapel Street. These were such an immediate success that only ten years later the Council obtained an Act to establish a new and larger cattle market in the parish of Walton and, in 1832, the original Stanley market was opened near the Old Swan village, about three miles from the town hall. By 1860, 3,300 cattle and 13,000 sheep were sold each week, about half by auction and half by salesmen, who often sold some by auction.

In 1882, there is reference to The Auction Yard, Breck Road where Thomas Carr was the auctioneer and officiated at the sale following the first Christmas Show of the Liverpool Cow-Keepers Association.

Shiploads of up to 500 cattle from America often arrived at Liverpool and Birkenhead but after the opening of the railway to Manchester, many imported cattle by-passed Liverpool market and went direct to that city. The introduction of refrigeration enabled beef and sheep to arrive as carcases and go direct to the meat, rather than the live, market. However, the live market continued to flourish with stock consigned from throughout Northern England, Scotland, Ireland and Wales. Reports in September 1880 record 2,135 cattle, 5,328 sheep ("small show in the market"). American cattle were reported as scarce but considerable numbers on their way. Prices for cattle were around £15 with sheep at £1 per head.

From the 1890s the importation of live animals to the Mersey had tended to concentrate across the river from the city at the Woodside lairage at Birkenhead, where large numbers came every week, in particular from Ireland. Great numbers were sold either by auction or privately with, of course, many consigned direct to other marts. In the autumn of 1935 some 2,500 cattle and 2,000 sheep were arriving most weeks.

Major dealers in the import cattle markets of the north west were E. Reed & Sons. Established in 1863, they had branches in Liverpool, Salford and Wakefield and, later, at Birkenhead. About 1875 the partners became alive to the need for some form of credit finance in the markets in which they dealt and they evolved a system of banking which, although quite unorthodox and dependent on trust, proved most successful and brought them good profits. The large importers and operators who collected the herds of cattle and sent them to the market for sale were enabled to draw at once the proceeds of each day's sales, leaving it to E. Reed and Sons to recoup themselves from the individual purchasers of the animals. A branch of this bank was opened at Woodside in 1911 and became the Cattle Trade Bank until it was acquired by the Liverpool Bank and Martins in the early 1920s and went on to become the most important of the Cattle Market

Branches operated by Martins Bank[14]. It closed in 1964.

Into the twentieth century, such was the level of the livestock business in the city that a larger, modern abattoir, meat and cattle market was built by the City of Liverpool and New Stanley Market was opened in 1931 at a cost of £690,000. It was the largest municipal undertaking of its kind in the United Kingdom and attracted major wholesalers including Borthwick`s who, in 1932, took a stall in the market and extended their Lancashire Cold Store to hold 3,500 tons of imported meat as well as expanding in Manchester and Leeds.

Stanley Market had accommodation for 2,250 cattle, 5,000 sheep and 2,500 pigs, but the improvements in the slaughter facilities encouraged a number of salesmen to sell on a deadweight basis, which is shown by the following weekly report of 24th October 1938 in the Liverpool Echo:

> *Monday – The week-end supplies were 291 cattle and 2,298 lambs, of which 1,092 lambs were sold by auction on Saturday. The cattle trade was slow. Choice small bullocks and heifers made 6¾d to 7d, heavy descriptions 6d to 6¼d. Young cows and bulls 5d, other classes around 4d. Scotch lambs were a good trade, but plain Yorkshire lambs were only in poor request. Scotch lambs made 9d to 10½d and plain Yorkshire lambs 8d to 9d per lb, sinking the offal. The livestock sent direct to the abattoirs for slaughter last week amounted to 1,186 cattle, 11,749 sheep and lambs, 498 calves and 1,239 pigs.*

After World War II and into the 1960`s, the motor car encouraged many city butchers, retail as well as wholesale, to go out to the country marts to buy their requirements direct from the farmer's local mart. This was to hasten the end of city auction marts and Stanley eventually closed in 1970.

Manchester & Salford

Manchester was a very large and dependable market[15] for fat stock from the North of England. Pens had been provided in 1774 on Salford Fairground and subsequently more pens were erected in Cross Lane with the Cattle Market there opening in 1837. The first Salford fat cattle show was held here in 1872.

Aided by the early railway from Liverpool Docks, it became one of the largest cattle markets in England where, as well as home bred

14 John Mashiter, Martins Bank Magazine, 1946.
15 John Bridge Newton is recorded as a livestock auctioneer in Manchester, holding sales c.1880.

stock, in many weeks from 1860 to 1890, some two thousand foreign beasts were sold in Salford Market, where Canadian cattle were always popular. In 1892 the annual Christmas Shows were started and among the promoters was James Nelson who particularly sold imported cattle.

Salford Cattle Market

He had lairages in London, Liverpool, Dublin and Barrow but his fortune was made when he employed Robert Galloway from Glasgow to go to Argentina to open a meat plant, known as Nelson's River Plate Meat Company. By the year 2000, the Galloway family owned Scotbeef and Scotlamb with modern abattoirs throughout Scotland.

In 1904, the Ship Canal reached Manchester at the huge cost of twenty million pounds. The Union Cold Storage Company built a large store for lamb carcases beside the market, which mostly came from Australasia and customarily destroyed the demand for lamb in local marts from mid-January until June.

In due course, a larger and grander market was developed by the City Council at nearby Mode Wheel. It opened daily from 7am to midday.

Tuesday was the main market day in Manchester when 3,000 cattle and 15,000 sheep were an average entry in its heyday. Cross Lane Mart closed in 1931 and the turnover at Mode Wheel declined after World War II. For the whole of 1963 only 3,206 cattle and 15,748 sheep were sold and this once great livestock market closed in 1964. A fine new dead meat market was opened but it steadily declined and is now demolished.

Outside Cross Lane Cattle Market, Salford

Manchester brings to mind two great characters who, in the 1930s came every Friday to my father's mart in Annan. Tom Jones from Manchester and William Riley from Liverpool, who, as well as helping the trade by buying 500 to 800 lambs each, also added to the good humour by insulting each other across the ring. My father would say that they did not bid too wildly against each other until the drink got into them but, even in that state, they shared the mart car, which took them back to Carlisle for their train.

Wakefield

In 1765 the Governors of the Charities of Wakefield, ten miles south of Leeds, purchased two acres of land near George Street to provide a market and at that time it was described as being a great agricultural centre.

The Governors let the site to six persons on a twenty-one year lease with these tenants agreeing to erect a sufficient number of pens for cattle and sheep and to keep the road and ground in good repair. This market was an immediate and continuing success with more land purchased in 1831 and 1836. Sales were held weekly, initially mainly by private bargain but with auction sales in the afternoon. The 1835 throughput, which paid tolls, was 258,000 sheep and 13,519 cattle. In 1860, 800 cattle, 4,500 sheep and 900 pigs were noted as being in the market on one specific day. By 1868, 360,000 sheep and 50,289 cattle were being sold annually and Wakefield was considered to be one of the largest auction centres in the country and with the best covered pens for pigs.

Wakefield, Cattle Market, 1930

There were weekly markets on a Wednesday and fairs in July and November. Mr Dodds was trading as an auctioneer from around 1870 and his firm, later Dodds and Robinson, held special sales of "Dairy and Pure-Bred Cattle" at a number of Yorkshire marts as well as conducting their regular Wakefield sales. Wakefield continued to be a leading centre until around 1930 when one of the key auctioneers moved away and, thereafter, the weekly sale seldom exceeded a hundred cattle and 400 sheep, whilst the pig market closed down. By 1960, M. Thomlinson and Son were carrying on the auction sales but weekly numbers were down to about 30 cattle, 150 sheep and maybe 130 pigs. These were small numbers and the writing was on the wall. The mart closed in 1963.

Before the auction system was established, market days, here and elsewhere, were often reported as being in a very unsettled state for some time. This often being caused by a few dealers or drovers with stock to sell "getting up a shout" that cattle or sheep were reported much dearer at a previous day's market by sending their men to run around the fairground telling this story and asking an extra shilling a stone for their stock. Not until the arrival of buyers who had actually been at the reported market who confirmed perhaps that prices had only risen marginally, or had fallen, was trading then able to get into its normal pattern. The afternoon auction sale would establish a fair price.

Leeds

The local auctioneers in Leeds seem to have been far-seeing entrepreneurs who chose to avoid the city centre congestion by having their marts a few miles out, close to suburban railway stations and with farms relatively near at hand.

By 1870, Messrs Lodge and Bartle had commenced auctioneering and, in that year, they held their Christmas Show at the Greyhound Inn. The band of the 5th Dragoon Guards were advertised as being in attendance and to be playing at intervals during the event. Surely an unusual distraction for an auction sale. By 1880, Lodge and Bartle had moved their mart out to Whitkirk, three miles east of the city and by 1900 they traded as Bartle and Son.

W. H. Pountney held his sales on the other side of the city, at Calverley Bridge, about four miles out but his office was at 14 Infirmary Street. He always advertised that his sales were "unreserved". A weekly Tuesday sale of livestock was held at the Swan Hotel, Leeds by Frederick Thirkhill. There was also a Leeds "Smithfield" market in North Street, named, as were a number of marts throughout England, after the famous London market.

Bartle`s continued to hold their sales at Whitkirk as it was gradually absorbed into the city proper and, by 1910, their average Monday sale might comprise some 40 new-calved cows, 150 fat cattle and 200 or so sheep. They also sold pigs although at that time the pig mart was closed owing to an outbreak of disease. Bartle`s office was in the neighbouring town of Garforth and they also held a Tuesday sale in the yard of the Greyhound Hotel, Vicar Lane.

After de-control in 1954 all Leeds sales seem to have been concentrated at Whitkirk where Bartle`s sold on a Monday whilst also selling at Selby, about eighteen miles distant, on a Friday. Numbers at Whitkirk were not much different in 1980 to what they were in 1910 and it closed in 2001 when the focus was on Selby, now one of the largest marts in Yorkshire.

Hull

Hull had been an important fair and market town since the fifteenth century. It was well placed for the import and export trade in cattle and sheep with Europe, across the North Sea, and this trade expanded greatly after the development of steam ships in the 1820s. As well as serving a prosperous agricultural hinterland, steam ferries across the river Humber also brought much stock from the rich farm lands of Lincolnshire.

The Cattle Market, Edwards Place, Hull
(Photo, Bernard Sharp)

To cater for these steam-age developments, a market evolved in Wellington Street and then, in 1838, fine new premises were built in Edwards Place where the auction system would later become the favoured method of sale. From 1878 to 1914 there were constant requests by the many auctioneers, led by Richard Marshall, that the mart should be extended and re-designed and this development finally happened in 1914. At this time, there were almost a score of auctioneers operating in the busy mart, selling some 800 cattle, 4,500 sheep and 900 pigs most weeks.

The market flourished and by 1939, there were the following twenty-four firms selling each week:

Richard Marshall
R. Hornsey & Sons
Robin Dunn
Clegg & Son
Leonard Nutchey
Tom Jackson
John W. Herd & Son
Ullyott & Wharram
R.R. Leonard & Son
Frank Hill & Leonard
Ernest Fisher
Evison & Evison
Megginson & Shipley
M. Featherstone
Frank Johnson
Frank Farnaby
J.W. Tuton
Tuton & Tuton
Tom Silvester
H. Jackson Executors
Cranswick & Cranswick
George Caley
Dee & Atkinson
Eric Wright

The auctioneers, as was customary, took turns at being first, second, third etc. with those with earlier turns generally attracting the larger share of the stock. Waiting six weeks to have an early turn must have ensured that some businesses were rather less than prosperous. Hull had four cattle rings in operation.

The Fatstock Control system introduced as war commenced in 1939 allocated every farmer to a local auction mart or collecting centre. This resulted in city marts, such as Hull, losing most of their traditional suppliers of livestock and a number of auctioneers took full-time jobs with the Ministry of Food as graders or clerks. Under wartime control, prime stock was graded and purchased by the government at fixed prices, which were lowest in September and rose every fortnight to a top in May. Sales were by weight so that farmers fed their stock to the greatest possible weight. The local auctioneer who handled the collections was paid on a headage basis.

During the war, Hull Mart suffered extreme damage from bombing and it was 1950 before it was rebuilt. In 1954 when freedom of sale was brought back and the auctioneers got back onto the rostrum, the city marts never regained their pre-war importance and many city auction firms did not re-open. Post-war, in Hull, the following eight firms recommenced:-

Hill & Leonard
G. Caley
R. Screeton
Tom Jackson & Sons
R. Hornsey & Son
Clegg & Wright
Dee & Atkinson
G.H. Evison & Sons

Large shows of prime cattle continued to be forward with, by the mid 1960`s, some 3,000 sheep each week. Hull remained one of the major pig markets in the UK with some 3,000 fat pigs in the pens on a Monday. However, as the years went by there proved to be insufficient custom and in 1982 a re-organisation left only the following companies in business:-

Frank Hill & Sons
R. Leonard & Son
G. Caley
Robin Screeton

Dunswell Auction Mart, Ralph Ward

They operated as Hull Livestock Mart Ltd until 1989 when re-development forced them to move to a new site at Freightliner Road in Hull, where the author visited them and was most favourably impressed by their new mart with its pens for 700 cattle, 3,000 sheep and 2,000 pigs. A new abattoir was also provided.

In 2002 the mart was forced, by various factors including an excessive rating assessment, to move to Skidby Ings

Farm, Dunswell, two miles north of Hull, trading as Frank Hill & Sons, Hull Cattle Mart, whose managing director, Ralph Ward has been a helpful source of information for this book. Primestock sales continue here weekly on a Monday with fortnightly store sales including pigs. Ralph`s first selling job was at Brigg and Louth in Lincolnshire and he works now, with Philip Mortimer and Charles Hill, to keep the last regular sales going in this area which once supported several large markets, selling thousands of cattle and pigs each week.

York and District

York was a trading city from the earliest times and its evolution serves as a typical example of the development of markets in major cities and towns elsewhere in this area and, indeed, throughout England.

In Roman times, Eboracum, now York, had seen a "gathering" of animals and produce to sell to the citizens within the occupied walls. The "Archbishop`s Fair" was held from as early as the first half of the 12$^{th}$ century when grants from his fair of St Peter`s Chains were made to the canons of York Minster and by medieval times, a number of markets were regularly held including "Thursday Market" and "Pavement". The former was where meat was sold. In the 16th century orders were frequently issued restricting the use of these markets to "foreigners" and "countrymen"; that is traders from outwith the city. Freemen of the city were expected to trade goods only from their own shops and "foreigners" could only stand in the designated market places. This regulation was only relaxed in special circumstances such as those in 1550 when, to improve meat supply during an outbreak of the plague, "foreign" butchers were allowed to sell elsewhere in the city rather than being restricted to selling in Thursday Market. Thursday Market, later St Sampson`s Square, was not restricted only to Thursdays but was generally held on Tuesday, Thursday and Saturday.

York, the old market with the famous twin cupolas

Meat was one of the main commodities sold in Thursday Market with orders about their sale made as early as the 14th century. From at least the late 15th century "foreign" butchers were restricted to the Market whilst citizen butchers had to sell their wares from their own shops, most of which were in the Shambles. In 1528, the country butchers were instructed to come to York on Tuesdays and Saturdays only. In the 18th and 19th century, Saturday was their day. The Minster clock would set the market hours.[16] Pigs and poultry were also regularly sold at this market.

16 Victoria County History of Yorkshire, volume re. York, 1955, provides much valuable information on the history of markets in York.

Postcard Souvenir of York Cattle Show, early 1900s

From the 14th century, a corn market was held in Micklegate but Pavement was the usual site for corn sales with trading commencing with the ringing of a bell. Initially All Saints bell, then Ouse Bridge bell which was superseded by the "Cornbell" in Pavement itself. Corn continued to be sold in the open streets "by sample" and a new corn exchange erected in 1868 was slow to attract farmers inside its doors, presumably as there was a cost involved, and closed in 1901. A new covered corn market was erected in 1926 between Coppergate and High Ousegate but sales continued in the street outside until at least the onset of World War II. In 1946 the corn market was moved to accommodation in the cattle market.

York, the New Livestock Centre, 1988
(left to right) John Burley, Senior Partner R M English; John Hopwood, Farmer; John McGregor, Minister for Agriculture; James Stephenson, Partner Stephenson & Son

Hay was sold for a time in the Shambles, which was also known, early on, as Haymongergate, but by the 18th century the hay market was held near Holy Trinity Church before moving, in 1827, to Peaseholme Green where a new weighing machine was provided. Wool markets were also held with the wool sheds being erected in the cattle market in 1862 to accommodate the sale of wool.

The York markets were exempted from Royal jurisdiction in 1316 and for much of the time thereafter the city benefitted from the privileges of the regulation of the markets. The corporation collected tolls and stall charges, although at certain periods the right to do so was let out to a third party, but by the early 19th century some tolls had lapsed and the collection of others was only negligently enforced. An Act of 1833 reaffirmed the corporation`s rights of regulation and collecting of tolls; in 1873, the market by-laws and rates of tolls were re-established and full control over all markets and fairs was confirmed to the corporation by the York Extension and Improvement Act of 1884.

In earlier times, there were a number of small cattle markets held in York with, for example, Toft Green being designated a cattle market as early as 1416. In 1502 there was a Royal grant of a fair to be held on the Monday after the Feast of the Ascension; lasting for six days, the first two days were devoted to livestock, cattle in Fishergate, horses outside Walmgate Bar and sheep on Heworth Moor. There was also a similar grant for a six day fair in the autumn on St Lukes Day (18th October); again two days were allocated to livestock but the locations used were different. By 1590, the city had received royal consent to hold regular cattle fairs on the Thursday before Palm Sunday and on alternate Thursdays between Palm Sunday and Christmas. Sheep were also sold from at least the 17th century. These fairs were held in Walmgate, Fossgate and the surrounding area. Later Long Close, inside the city walls from Fishergate to Walmgate Bars was also used. Meanwhile cattle sales continued in the streets and caused considerable issues until in 1827 the corporation constructed pens and established a cattle market on six acres of land outside Fishergate Bar. The site was improved by the erection of the City Arms Inn in 1828; it proved to be a famous hostelry for farmers and dealers for many generations.

The fortnightly fairs were held in the market but the fair on the Thursday before Palm Sunday continued to be held as before but died out well before the end of the 19th century. The first "Fortnight Fair" to be held in the new premises brought in tolls of £18 and a profit of £10 which was considered to be a very satisfactory start. Thirty two fairs were held in the first year of operation and brought in £352 of revenue. Special fairs were held at Candlemas (2nd February) and a cattle and horse fair was held on All Souls Day (2nd November). The date of this was later moved slightly and in due course, from 1827, the cattle were sold in the cattle market. In 1841, the Yorkshire Gazette reported that the horse fair took place "as usual" outside Walmgate Bar, with cattle in the new cattle market.

York, the main cattle ring, 2013

Horses had, of course, been sold from time immemorial and three of the oldest horse fairs were appropriately held at the "Horsefair", an area outside the city walls at the northern end of Gillygate. In 1587, the horse fair was ordered to be held as usual "according to the ancient custom". Horse fairs were still being held in 1818 but by the late 1820s had been merged into the new cattle market complex.

In the 16th century, pigs were marketed at Peaseholme Green and also sold in Swinegate followed by the establishment in 1605 of a swine market outside Fishergate Bar. In 1738, Toft Green was appointed as the place where pigs were to be sold and these were sold there on a Wednesday until about 1818. In the 19th century, a pig market was also held near Foss Bridge before pig sales were also absorbed into the cattle market.

The cattle market was reconstructed in 1855 and a fatstock show instituted in 1857. The wool sheds were added in 1862 and extended in 1876. Cement floors were laid to the market pens between 1879 and 1882 and a railway branch line and siding was built. Land was bought in around 1898 for an extension to the market. Improvements including the addition of two auction rings, with their distinctive twin cupolas, were made in 1904. Following tradition, markets continued to be held fortnightly, with weekly sheep sales in the autumn, until 1914. Afterwards sales were weekly until the mid 1920s when Mondays and the traditional Thursdays became regular sale days and the corporation provided additional space which, aided by the closure of certain streets on sale day, enabled this city centre market to prosper until its closure in 1970.

As the 19th century progressed, the old tradition of sales by private bargain had been gradually superseded by auction sales. Thomas Walker was selling horses by auction in York in 1860 and, by about 1870, he expanded into cattle and sheep sales, followed soon by other auctioneers including Roger Iddison.

The Stephenson family were amongst the first auctioneers in the York area and had been involved as valuers from as early as the 17th century, acting for the Fairfax family at the end of the English Civil War in 1645.[17] In 1871 Jacob Stephenson had commenced selling in Tadcaster Mart, less than ten miles from the city; and he started sales in York, probably at the Wool Sheds, in about 1881. By 1900, Stephensons were paying the tolls for the pennage in York along with seven or eight other auctioneers.

York went from strength to strength as an auction centre and the corporation provided additional space in the 1920s. Eight firms of auctioneers were still operating the market at the outbreak of war in 1939 when government control took over all prime stock sales.

The meat shortage and consequent control continued long after the war finished in 1945 and it was not until 1954 that the Government finally released its restrictions and allowed auctions to start up again. At this time, only four firms took up their gavels again in the York rostrums – R.M. English of Pocklington, a large professional business, which had already diversified into selling grain and was dominant

17 For more on the Stephenson family see Chapter 13.

in the East Riding of Yorkshire; Joseph Cundall & Sons of Sherburn, whose roots were in Ryedale, north east of the city, where they also operated at Malton Market; Don Johnston, an independent auctioneer whose brother, Basil ran the Boroughbridge Mart; and Stephenson & Son, where the senior partner, Reg had three young sons in their teens.

The area around York is topographically pretty flat and within the Vale the confluences of many rivers meet, including the Ouse, Wharfe, Nidd, Derwent and others. The riparian countryside was subject to seasonal flooding, but provided good summer grazing, which was excellent for fattening store cattle as long as they had enough age to make use of the herbage.

Ireland was full of such cattle, from two to four years old, and large numbers crossed St George's Channel to Holyhead for the onward train journey to Layerthorpe Station at York, where they were unloaded and walked out of the city to lairage on a Corporation owned farm at Osbaldwick. Often 1,000 head a week arrived by special trains. Every Thursday morning, rested up, the cattle would be walked back to the market which, by this time, had grown to accommodate 2,000 head including pennage directly beneath the walls between Walmgate Bar and the Fishergate postern.

The Irish cattle dealers used to travel over with their stock and pay tolls to the Market Superintendent to sell them privately to farmer customers. James Stephenson, the middle of Reg's sons, remembers buying his first pen of 25 Angus X Irish bullocks for £48 in 1959. The auctioneers never seriously resented the private dealing of the Irishmen as they provided an inexhaustible supply of fatstock for selling in the months to come. In any case, alongside the Irish trade was homebred stock to sell and a nascent dairy market was set up to accommodate the growing number of cow herds on the mixed farms around York.

An intensive national TB testing programme created a premium for clean stock and an "Attested" market was started at York, selling up to 100 dairy animals per week with such major dealers as Albert Hall and Mr Lupton carving out a living for themselves.

The general transition in farming practices, particularly in the East Riding, led to a steady loss of grazing to an arable regime and the discontinuing of the traditional practice of arable farms feeding large numbers of cattle on straw in sheds; not to mention the falling away of the traditional fattening of sheep folded on root crops. The Irish cattle trade died away and livestock numbers were greatly diminished. This was, inevitably, a big loss to York and many other Yorkshire marts.

By the 1960s the city fathers were focusing on reconstructing their urban environment and Lord Esher prepared a grand report for York's centre and the view was taken that, for various reasons, not least traffic congestion, the mart should be moved out of the city. From this flowed the plans to relocate the market and the corporation abattoir right away from the city and, to this end, a twenty-two acre field was purchased at Murton, 4 miles from the Minster. Seven acres of the field would be devoted to building the new abattoir in which the farmer cooperative, FMC, was heavily involved; and architects drew up designs for a cattle market, corn market and a two hundred bedroom hotel on the remaining 15 acres, but suitable prospective tenants couldn`t be found to take the leases. In 1968/9, the City Council tendered the sale of the site on condition that a market was built to their approval.

Reg Stephenson sounded out his fellow auctioneers, but only R.M. English wanted to join in and then with a minority share. Their tender was initially rejected in favour of a larger bid by Duncan Davidson, who later went on to create the Persimmon house building empire. However, Mr Davidson could not find a lessee and handed back his offer, which left the way open for the Stephenson and English partnership to build the new York Livestock Centre.

The work took 12 months to complete and cost around £200,000 and the market was opened by James Prior, Minister of Agriculture, on the 1st March 1971. Stephenson had a three-quarters share and English a one-quarter share of the market, which was extremely well designed and built to a high standard.

Like many marts at that time, pigs were an important area of the business, which was, in due course, greatly damaged by the Fatstock Marketing Corporation, who bought up many bacon factories and generally refused to buy in the marts. That part of York Mart has been replaced by a social area, which one hopes is as profitable as the pigs once were.

The York Livestock Centre continues to operate very successfully and its story would not be complete without mention of the massive York machinery sales when over 5,000 lots are sold through five or six rostrums, each rostrum being manned by two auctioneers taking turns. Usually one auctioneer is from York Livestock Centre and the other from an outside firm coming in to help out.

An innovative feature at York is the loyalty card scheme which rewards sellers with points which are then cashable for many benefits including either buying meat at a discount from butchers who buy at the mart or paying for tickets for the Yorkshire Show or some other desirable event.

A number of other firms and marts within striking distance of York thrived for many years. When markets were decontrolled in 1954 and allowed once again to sell via live auction, there were thirteen markets within twenty miles of York Minster – Tadcaster, Boroughbridge, Driffield, Easingwold, Goole, Knaresborough, Malton, Pannal, Ripon, Selby, Thirsk, Wetherby and York itself. Of these only York, Selby, Thirsk and Malton remain.

The Stephensons had been succeeded at Tadcaster Mart by M. Thomlinson & Sons who also sold at Wetherby Mart until it closed in about 2001. They had also held sales at the Wool Sheds in York.

Boulton and Cooper, an old established business, managed by Brian Wilkinson and. latterly, Philip Place had weekly mart sales in Malton, Pickering and Seamer; Seamer closing in 1972 and Pickering in about 1980. The firm was acquired by Stephenson & Sons in 1987 and this, later, led to them amalgamating in 2010 with the mart business of the large firm of Joseph Cundall and Sons, now managed by Derek Watson. Cundall`s, whose head office was for many years at Sherburn, between Malton and Scarborough, sold principally at Malton, where they and Boulton & Cooper took turn about, alternating being first sellers for cattle and first for sheep in each week. Cundall`s had also traded in York every week until the closure of the old premises in 1970. They also held a few autumn village and moorland sales of store and breeding sheep and ran Kirkbymoorside Mart under the Kendrew and Dennis banner. Malton Mart continues today and is very traditional, with sheep pens erected in the streets. It would benefit from replacement but the cost of a modern mart is perhaps not an economic project at this present time.

Joseph Cundall`s grandsons, John and Harry, were held in high esteem and conducted a most successful business. My elder son, John, worked with them as a trainee for a time. John Cundall was a tremendous auctioneer who would sell 1,000 fat cattle on his own without a break. Their staff and supporters were always sorry that the Cundalls took the decision not to participate in the new York Livestock Centre.

In marts owned by the local council, the auctioneers took turns to sell the stock consigned to them. The order was either fixed by drawing lots or by rotas agreed by the firms for the forthcoming year. A good place in the ballot was essential as farmers usually supported those with an early slot as only a very few top auctioneers could depend on support regardless of their position in the sale. Most auctioneers advised their friends to send stock only when they were first or second in the selling order.

The following advertisements from 1960 explain the system in Beverley and Hull, which was a similar pattern to marts throughout the country from Newcastle upon Tyne to Gloucester in the south. The tremendous regular entry of prime cattle is notable but all are now gone from the marts.

From the Hull Times 3rd September 1960

EAST RIDING CATTLE MARKETS

ENTRIES OF FATSTOCK AT THE FOLLOWING MARKETS

	Pigs	Cattle	Sheep
Monday Sept 5 – Beverley 10am	49	205	90
Monday Sept 5 – Hull 12.30pm	256	1,320	590
Tuesday Sept 6 – Driffield 10.30am	61	407	81

BEVERLEY CATTLE MARKET

Auctioneers' Turns for Fatstock, Monday September 12th

	Pigs	Cattle	Sheep
Dee & Atkinson	No. 1	No. 6	No. 4
G.H. Evison & Son	No. 2	No. 7	No. 5
R. Hornsey & Sons	No. 4	No. 2	No. 7
Tom Jackson & son	No. 6	No. 4	No. 2
Leonard & Hill	No. 7	No. 5	No. 3
F.W. Ullyott & Sons	No. 3	No. 1	No. 6
Eric Wright	No. 5	No. 3	No. 1

HULL CATTLE MARKET

Auctioneers' Turns for Fatstock, Monday September 12th

CATTLE

1. George Caley
2. R. Hornsey & Sons
3. G.H. Evison & Son
4. Eric Wright
5. Dee & Atkinson
6. Tom Jackson & Son
7. Hill & Leonard
8. Clegg & Son

SHEEP

1. Hill & Leonard
2. Clegg & Son
3. George Caley
4. R. Hornsey & Son
5. G.H. Evison & Son
6. Eric Wright
7. Dee & Atkinson
8. Tom Jackson & Son

PIGS

1. G.H. Evison & Son
2. Eric Wright
3. Dee & Atkinson
4. Tom Jackson & Son
5. Hill & Leonard
6. Clegg & Son
7. George Caley
8. R. Hornsey & Sons

CHAPTER VI
HEALTH & TRANSPORT

In 1745 the cattle trade of Britain was seriously disrupted by the worst ever recorded outbreak of the cattle plague, Rinderpest. This came from Europe and, possibly, because it coincided with Bonnie Prince Charlie`s invasion from Scotland, the government were slow to react. By 1748, the disease was raging in Yorkshire and Lancashire and the next year it reached Settle, by which time all fairs and markets were closed and a night curfew on movement of all cattle was imposed.

This was the first time that the government had restricted movement of stock and paid compensation for destroyed animals but, nonetheless, it was ten more years before the disease would be brought to a halt. Disease control would be greatly improved over the succeeding half century as the movement from common grazing to enclosed fields gathered pace. Control was also helped greatly as farming became the hobby of the wealthy who had the inclination and funds to make continuous improvements in farming practices. They were particularly supported by King George III, "Farmer George", who personally campaigned against the importation of cattle and hides from infected European countries. Indeed it became customary for the King to open Parliament with a personal report on the cattle disease and a special prayer was said in churches in a further effort to combat the pestilence.

The first veterinary college was established in London in 1791 and, two years later, a Board of Agriculture was established with Arthur Young as secretary and some government finance. It was highly influential but only lasted until 1822 when a financial crisis forced its closure, just as it was investigating the disease of sheep scab and the need for regulation to mitigate cruelty and to enhance animal welfare.

One result of the restrictions emanating from the 1745 disease outbreak was that the pattern of trade changed as weekly markets expanded with local livestock salesmen taking much of the trade from the more regional seasonal fairs. These local market salesmen would, in due course, be replaced by the auction marts of the nineteenth century but, meanwhile, many prospered with their finances improving when they were appointed government buyers for cattle, sheep and pigs to feed the large navy and army of the Napoleonic wars. During these wars, up to 1815, these agents became quite wealthy but were not in the same league as their superiors, the designated supply merchants, many of whom became millionaires rewarded by Knighthoods.

The development of steam ships from the 1820s onwards brought speed and regularity to the supply of livestock. Aberdeen, Newcastle and Hull Steam Navigation Co., had weekly sailings carrying livestock and, as late as the beginning of the twentieth century, a return journey to Hull from Aberdeen by a cattle dealer was advertised as only costing £1, but the fares for his cattle and sheep were not stated.

Livestock imports had been small until this time but steam power ensured that they rose rapidly so that by 1849 some 75,000 cattle and 140,000 sheep arrived from Europe. A single unloading wharf for London was established at Deptford, which had been a large Naval dockyard. There was also a substantial livestock shipping trade between UK harbours and ports; sailing often being more convenient than overland travel. For example the small harbour of my home town of Annan in Dumfriesshire, Southern Scotland, had two sailings per week from the Solway to Liverpool and in 1846 the Harbour Trustees reported that 2,160 cattle and 70,400 sheep had been shipped to Liverpool that year.

From the late 18th century, canal transport had also been a crucial benefit to the agricultural industry and the 1816 opening of the Leeds and Liverpool canal had changed business life in a significant part of Yorkshire and Lancashire, providing easy transport for twenty ton loads to glide along the bright shining water between the shadows of the many mills that lined the route; and pulled only by a single horse. The canals carried the farmers` produce to market and the surplus feed and fodder from one area to a district that needed it more. Show cattle were sometimes carried and, often, pigs sailed to market.

As in 1745, the ready import of livestock from Europe continued to have its drawbacks in terms of health particularly as infected animals could be quickly transferred from ship to train, without really being properly observed, and then be distributed across the country. In 1865, imported cattle started an immense outbreak of Foot and Mouth disease which, for the first time, was recognised in sheep.

The disease was believed to have arrived at Hull dock from Europe and lasted for two years with the loss of some 400,000 cattle. The weekly

Farmer newspaper reported that the severe cattle plague had arrived in London and by June many city dairies had been killed out with the result that milk was almost unobtainable and the price doubled to four pence a quart. The House of Lords set up an inquiry which reported that in 1864 159,000 cattle had been infected in Russia, of which 105,000 died.

On the second of August, The Farmer newspaper reported that 14 cattle were sent from London docks to Kelso for sale, four days later one was ill and died two days later, in all thirteen died and other cattle on the farm showed infection so that it was decided to slaughter all the farm cattle. Nonetheless at the same time the Highland Show was being held in Inverness, with stock present from all over the country. The next week the government advised farmers and dealers to keep newly bought stock separate for a period of 12 to 14 days. There would be immediate slaughter of cattle infected, which was known by discharge of watery fluid from eyes, nose and throat. Hides were retained by the farmer in vats of disinfectant to sell later.

Market premises were ordered to disinfect thoroughly with Chloride of Lime, but the Falkirk Tryst went ahead. Thereafter, markets and marts were inspected continually by a Government Veterinary Department which was established in 1869.

In September 1865 it had been stipulated "That no proprietor of cattle expose any animal in any public market for the next three months without a certificate from the inspector. If so exposed without a certificate, he shall be liable to a fine of not less than 50s., and not more than £5, for every head of stock so exposed; and if any animal affected with the disease be exposed, it shall be liable to be destroyed, and the owner fined in a sum of not less than £20, and not more than £50".

The Yorkshire Agricultural Society took the matter most seriously with an extraordinary meeting of the members being held at York to consider the best means for preventing the spread of the cattle disease. Admiral Duncombe, M.P., occupied the chair. Colonel Leeman said the North-Eastern Railway had taken every means to prevent the introduction of the plague, and that any of their officers would be dismissed if found lax in enforcing the regulations. Resolutions were passed recommending farmers and agriculturalists to abstain from purchasing cattle in the markets. If such purchase were necessary, however, purchasers were recommended to place such cattle in quarantine for fourteen days. If disease symptoms appeared, the cattle affected should immediately be inspected by a properly qualified person and, if necessary, the cattle destroyed, and buried at considerable depth. Constant cleansing of yards and premises was also recommended. The Council further suggested that inspectors for petty sessional divisions should be appointed by the Justices, to act on the first appearance of the disease, and also the establishment of cattle insurance associations.

The government provided for the payment of inspectors by a letter from Whitehall:

"Sir, - I am directed by Secretary Sir George Grey to acknowledge the receipt of your letter and to inform you that the Lords Commissioners of Her Majesty's Treasury will sanction the payment to inspectors, appointed under authority of the order, in districts infected with the 'Cattle Plague' of the sum of £1, 1s. for each day that they are actually engaged in the performance of the duties required of them by the said order, with travelling expenses at a rate not exceeding 3d. a mile each way."

The inspectors` duties included:

Every such inspector shall have power within his district to direct that any animal which he suspects to be labouring under such disease shall be kept separate and apart from every other animal which appears to be free from such disease. And every person having the charge of any animals shall obey every such order or direction given.

Every inspector shall have power within his district to seize and slaughter, and to cause being buried in any convenient place, any animal labouring under such disease.

Every inspector shall have power within his district to cause to be disinfected, or, if necessary, destroyed any fodder or manure.

That is shall not be lawful to send or bring to any fair or market, or to send or carry by any railway, or by any ship or vessel coastwise, or to place upon or drive along any highway, or the sides thereof, any animal labouring under such disease.

Every person offending against this order shall for every such offence forfeit any sum not exceeding £20, which will be payable to the Justice before whom he or she shall be convicted of such offence.

And in a letter to the railway companies, the government wrote:

"Privy Council Office, Whitehall, August 18.

"Sir, - I am directed by the Lords of the Council to request that you will be so good as to call the early attention of the directors of your company to the enclosed copies of four orders which have been issued by their Lordships, with a view to prevent the spreading of the cattle plague. Their Lordships are persuaded that your Board will not fail to recognise the magnitude of the mischief involved in the rapid extension of this very serious disease, and that their Lordships may rely on your co-operation."

In 1872, a large outbreak of Cattle Plague in Yorkshire again originated in imported cattle.

Contagious diseases in livestock were a serious problem for farmers, dealers and auction marts with no public authority to deal with them until an Act of 1878 gave local authorities power to slaughter infected animals and pay compensation from local rates. It stated that compensation for the destruction of infected animals should be at the rate of one half the value of a healthy beast but that animals only suspected of the disease should receive their full market value prior to slaughter up to a maximum of £40 per head from local rates. This introduced auctioneers as official valuers for the first time in England although, previously in Scotland, Mr John Swan, the Edinburgh cattle salesman had been appointed to advise the local authorities on such matters.

As one might expect, each local council had its own interpretation of the act and pursued its own policy, or lack of policy, but a new act of 1890 proved to be a marked improvement as the national government took over the power of slaughter and the payment of compensation became the responsibility of the Board of Agriculture, which had been re-established in 1889 and who were ordered to eradicate Pleuro-Pneumonia and other diseases from the national herd. In 1887 there had been around four hundred outbreaks of Pleuro disease in Britain with standstill orders in fifteen counties, including Lancashire.

Among other things, was enforced the cleansing of marts where water supplies were hurriedly provided, although cleansing of marts selling Irish cattle had already been introduced in the previous year.

By this time there were over ten million cattle and thirty million sheep in the country and the government were taking a keen interest in auction marts, both in respect of their internal operation and in regard to their effect on the neighbouring community. The Board, from their office in Whitehall Place, London and their inspectors were in touch with many local authorities giving both advice and direction.

In particular, they wanted to be involved in the many schemes to get livestock off the market squares and streets and hotel yards and into more spacious penned marts as near the railway as possible.

The following letter from the Secretary of the Board to Warrington Town Council reveals their objectives:

Board of Agriculture
4 Whitehall Place
London
6th May 1895

No. 635/95
Sir,
I am directed by the Board of Agriculture to advert to your letter of the 8th inst., respecting the incorporation of the existing Market, Fair and Saleyards in the borough of Warrington, in the proposed new Public Cattle Market, and I am to forward to you for the information of the Local Authority the enclosed Extract from a Report received from one of the Board's Inspectors, from which it would appear that the site upon which it is proposed to establish the new Mart is a very suitable one.

The Board would be glad to be informed as to what decision the Local Authority may arrive at in the matter.

I am, Sir, Your Obedient Servant
T.H. Elliott
Secretary

Extract

I have this day met the Mayor and Town Clerk of Warrington and inspected the site alluded to in the latter's letter of the 8th instant. The site is the yard of the old Militia Stores and appears to be a very suitable place, close to the L and NER Station and might easily be converted into a cattle market, the drill sheds which occupy a portion of two sides, would when paved, be fitted for cattle standings. At present the market arrangements and cattle sales are very unsatisfactory and quite unworthy of such an influential town as Warrington, the market being held in the streets, the sales in yards of Hotels. It being essential that some improvement should take place, the proposed site, is certainly much to be preferred to that of the old fairs ground.

Chairman

In 1919, the Board of Agriculture became the Ministry of Agriculture and Fisheries, subsequently Agriculture, Fisheries and Food. It`s responsibilities now come under the auspices of D.E.F.R.A, perhaps otherwise known as the Department of Interference and Environment.

Minor disease outbreaks, requiring the attention of the Board and Ministry, continued throughout the years until 1967 when there was, again, a massive outbreak after which meat imports were banned from countries where Foot and Mouth disease existed. Furthermore, swill for pigs had to be cooked; however, this was not always enforced with, of course, devastating consequences in 2001.

The 1967 outbreak was followed by a very full inquiry led by the Duke of Northumberland which issued an excellent report among whose recommendations was the control or exclusions of imports of meat from countries where Foot and Mouth existed, particularly from the Argentine. They also suggested regular reviews of administration such as valuers' fees but unfortunately this matter was ignored and a battle royal ensued in the 2001 outbreak when auctioneers rendered their accounts.

In 1970, a scheme for the compulsory eradication of Brucellosis in cattle was introduced. This involved marts in considerable trouble and expenditure in providing segregated premises for the sale of accredited animals. Similar issues had been encountered in much earlier schemes for the eradication of Tuberculosis and were later to occur, on a much smaller scale, with the accreditation of sheep which had been screened for Maedi Visna disease. BVD is now the health problem to be eradicated and, whilst certainly a desirable programme, yet again the marts are involved in additional record keeping and administration.

It was often said that it was the railways that made the marts and certainly the city markets and salesmen started to depend on railway deliveries from the 1840s. However, more rural lines were built later, from 1850 to 1880, and it is difficult to know which came first; the railway providing new markets for agricultural producers or the new marts providing a compelling reason to build an uneconomic railway. Whatever the truth, it probably resulted in both too many marts and too many small stations. The stations were closed by Dr Beeching and the marts by the supermarket chains. In 1932 almost twelve million livestock were moved by rail, declining by 1962 to only one million with the result that in 1964 livestock facilities were withdrawn from all small railway stations by the British Transport Commission. Rail had been a useful and convenient service for over a century but succumbed to the convenience of the lorry, which came right into the farmyard. Thirty years later lorries were fewer and a Landrover with trailer took most of the livestock to market.

Ten years later all livestock services were withdrawn from the railways and the age-old sheep mart cry "station" was heard no more. As the auctioneer called the name of the buyer, he would have replied "station" and the seller's shepherd would immediately drive his sheep off to the station clutching a loading card given out by the auctioneer's clerk. This card conveyed the number, name and destination to the shunter, who would then load the trucks in the right order with, say, fifty to seventy lambs in each, according to their size and positioned according to their direction of travel. Since 1880 marts advertising "Stock Trucked Free" had been particularly popular with buyers.

The traditional single pen at the village station had three gates fronting the railway so that a shunter in a hurry could scarcely fail to leave the wagon in a loadable position. Railway wagons were all the same size, each holding fifty Suffolk cross lambs or a dozen prime cattle. However, they were charged as large, medium or small and the charge was effected by moving a heavy partition either three feet or six feet up the wagon where it was hung by a slot at each side. It was always good to persuade the railway official to let your stock go large for the lesser price of a medium or small. This saved the farmer expense and saved the railway man from some very heavy work.

I had often helped unload sheep from the railway wagons in the evening and found that the stock ran easily in the dark. We unloaded all our wagons into the pens then I was sent two hundred yards up the road to open the field gate after I shouted "ready" our mart yardsman would open all the pen gates when the sheep ran briskly into the field. So at the age of ten I was becoming a keen drover, although the excitement of sitting in the signal box while waiting for the train of goods wagons may have encouraged me. Goods trains did not run to an exact timetable as they had to be shunted into loops or sidings to let passenger trains fly past to keep up to time.

From 1860 refrigeration began to be developed and would grow in ten years to allow the refrigeration of a whole cargo of meat, which enabled ships to bring frozen meat to the butchers of Britain. Thus began the cold chain of refrigerated ships, trucks, shops and, in about 1910, kitchen refrigerators that would abolish the farmer's monopoly of fresh food supplies. Things would never be the same again as aeroplanes added to this chain that undermined our markets and made possible the rise of supermarkets drawing supplies from every corner of the globe.

Unloading cattle, a typical scene at the railway sidings, this one at Rugby

CHAPTER VII
PROGRESS UP TO THE GREAT WAR

In the nineteenth century many local agricultural groups were formed in order to promote the knowledge of agriculture and the success of the industry. Many of these groups established the local or county show and some of these continue today, including the wonderfully successful Great Yorkshire Show, which recently held its 160th event. Some groups evolved into Farmers` Mart companies whilst others restricted themselves to discussions, farm visits and an annual dinner. In Yorkshire, some of the groups lightened the occasion of their annual dinner with the annual award of the White Hat. This was given for the most untidy stack yard, with particularly poorly thatched stacks, in the district and there would be great glee and mirth if it was won by a member of the committee.

Farming in Britain enjoyed about forty years of continuous prosperity from 1835 to 1878. This came to an end largely owing to the development and growth of refrigeration and the subsequent massive increase in imports[18] and was marked by the collapse of the City of Glasgow Bank and many other local banks as the production of virgin lands in the Americas and Australasia reached our cities. Add to this, three years of extremely bad weather and, by 1880, farmers were certainly carefully studying their sales pattern and, to achieve the best prices possible, they turned in droves to the auction marts.

The marts expanded rapidly and soon advertisements would name the "Old Mart" and the "New Mart" in many towns, where the new mart was often set up by a small band of farmers as a "Farmers` Mart". In these cases, the farmers would each invest in a few shares and the business would be run by a board of directors employing a suitable auctioneer. In my experience, all marts had to be operated as a farmers` mart or they did not survive. They had to put the needs of the farmers and customers ahead of that of the owners. This still holds true today.

The directors of Farmers` Auction Companies were often guilty of reducing the commission charged to a level that made it impossible for either the mart or their staff to flourish whilst also undermining their more prudent owner-operated competitors. In many marts they

18 See Borthwick's "A Century in the Meat Trade", they had started in Edinburgh, Liverpool and Manchester before spreading to Australasia.

charged two shillings or two shillings and sixpence (12½ new pence) to sell a dairy beast. This would be less than one per cent of the price, which was very generous to the sellers and did not enable the mart to build up any capital reserve.

The auction marts were helped greatly by the large amount of free publicity given to them by the local and regional newspapers. These publications attracted many new country readers who either wanted to see their name in print in the sale reports or needed to know how prices were going; something that had never been factually known before. The free reports were offset by the costly advertising, which ate into the small profit of the local mart. Some auctioneers were extravagant, placing self-promoting large advertisements or repetitive insertions which were, of course, outdone by their neighbours. The City mart salesmen often agreed that they would all use a similar small advertisement and that saved them a lot of money and also reduced the ill-will caused by exaggerated claims - although it was remarkable how many auctioneers sharing a ring could claim to be in the best position in the ballot!

Among local newspapers that prospered on the back of farming and the auction marts was the Preston Guardian, which was first printed in 1844. It continued to support farming and now comes to us as The Farmer's Guardian, which covers the entire north of England with every auctioneer advertising in it. A Scottish edition has recently been launched, which always proclaims that "The Mart is the Heart". Among the traditional Scottish farming papers, the Scottish Farmer alone survives and continues to be successful, carrying detailed coverage of auction sales and extensive auctioneers` advertising.

Livestock prices continued to fall in the latter part of the 19th century with a consequent fall in mart incomes. By around 1890 many auctioneers realised that they couldn`t go on as they were and must either retire or join together with another auctioneer to provide a more viable mart. It was a decade of consolidation, often achieved by allowing each of the individual auctioneers to continue holding sales on farms on their own account whilst acting for their auction company, of which they may even have been manager, in respect of sales held in the mart. A particularly successful example of this was Richard Turner at Bentham who continued to conduct annual village sales and farm sales over a wide area on his own account rather than for the Bentham Mart company. Mr Taylor of Skipton also continued to hold village sales two or three times a year at Malham and Kilsey on his own and not for the Skipton Mart.

Throughout the latter part of the nineteenth century, the weighing of fat cattle in marts had been a seriously divisive issue with a few farmers in favour and the butchers and dealers fiercely opposed, as their greater experience gave them a decided advantage. Westley Richards of Ashwell Lodge, Oakham, Rutland, writing in 1891, estimated the loss as at least £2 per fat beast or over ten per cent of its value by underestimations of weight[19].

In 1881 John Swan & Sons installed a weighbridge at their mart in Edinburgh and in 1889, the government passed an Act to encourage weighing by the Markets and Fairs (Weighing of Cattle) Act, which was not always applied. However, the pressure mounted and in January 1890, Thomas Corson started to weigh and record these weights at his mart in Stirling.

Auctioneers in Yorkshire, Lancashire and Cheshire followed the example and it soon became almost a universal practice with the weighbridge for a single animal costing around £10. A considerable investment for some auctioneers, who claimed that it cost another £10 to install it.

Weighing of fat sheep became the general practice with the Second World War and has continued ever since. Surprisingly Mr Richards recorded that sheep had been the first animals to be frequently weighed with Yorkshire leading the way. In the 1850s it appears that many thousands of sheep were sold by liveweight for the Leeds and Wakefield markets, but somehow or other the weighing was allowed to fall into disuse. Probably the butchers did not like it.

The issue of the weighing of store cattle remains unresolved with our mart at Dumfries in Scotland weighing all store cattle whereas, just over the border, at Longtown, the store cattle are still not weighed.

In the last quarter of the nineteenth century many marts removed from town squares and the yards of coaching inns to better sites, often near the railway station. These sites served well until the Land Rover and trailer necessitated another move to larger, more accessible sites with big areas for parking lorries, trailers and cars as well as the restaurants, pubs, shops and offices needed to cater for the customers.

The fairs had finally disappeared as a competitor to the auction system and, for many auctioneers, the most important days of the year were those of the Christmas Show and Sale weeks when three different

19 Highland & Agricultural Society Journal, 1891.

days might be allocated to cattle, sheep and pigs in turn. My late colleague, Graham Laurie would recall that, prior to the war in 1939, his father would be on the road for almost a month canvassing more stock than the mart could hold and then begging the various buyers to be there on a special day which did not conform to their usual pattern of attendance.

On some occasions, police were required to control the crowds of the interested and merely curious who packed the marts at these events when local butchers bid well to buy the best for the housewives who supported their shop. These Christmas sales are still an important day in the calendar of most marts.

No Christmas Show day was complete without a dinner, either at lunchtime or after the sale, at which a very large company was always reported. Some auction companies contributed one shilling a head towards the cost and amongst the many toasts was one to the auctioneers, usually welcoming the payment of a small dividend. The auctioneer in reply always said how well the mart was doing and how much he appreciated the support of all those present that night, which had led to the great success of the mart. "In his work he had tried to impart a spirit of fairness and justice between man and man, which would always bring reward to an auctioneer", so said the great, Northumbrian and Yorkshire auctioneer, T.T. Iveson at one such event. Toasts were proposed to the judges, the prize winners, the buyers, the town and trade with a few unexpected toasts interjected by the jolly diners. So an enjoyable day would come to a close with ponies and traps carrying their well-entertained occupants home.

During the Great War, in 1915, the government asked livestock auction companies to allow recruiting officers to address audiences at their sales. Most of them agreed. Throughout the conflict, marts and auctioneers were involved with the conduct of the huge free gift sales that took place to raise money for war charities, always giving freely of their time and effort as well as buying unwanted lots. The whole landed and rural community entered into the spirit of these sales and large sums were raised. This charitable tradition continues today and, in early 2019, the auction marts of Cumbria held a joint auction which raised £58,000 for charities and good causes.

The early years of the twentieth century saw a few more marts opened. However, the war of 1914 put an end to some of the old ways as young auctioneers, their clerks and mart men joined Kitchener's army with many of them killed or wounded in action causing the premature demise of a number of established auction businesses. Around 3,000 members of the Auctioneer's Institute for England and Wales joined the Army and the Navy and it was estimated that over 8,000 men in total from auction mart firms joined up.

The recruits found, on joining up, that war had taken on a new dimension. It was fought by the internal combustion engine, not only tanks on the land but submarines in the sea and aeroplanes in the sky. These made it a more fatal encounter and many did not return.

In 1916 the German submarines were causing food shortages and by 1917 the marts were grading and buying slaughter stock for the army and navy.

There were rumours of competitive grading between auction firms in order to attract business and so, in 1918, every farm was allocated to its nearest mart and the system then worked very well although to the detriment of the City marts. The auctioneer was paid 10/- (50p new money) for every beast and 1/6d (7½ new pence) for every sheep, which was a fair reward for their services.

1917 was a year of serious and justified worry regarding the scarcity of food with there being real fears of starvation and the government food controller preparing for a general rationing scheme. In the marts, the number of grades for cattle caused much debate with auctioneers wanting the simplicity of fewer grades, while the farmers wanted more grades so that prices would reflect that higher quality which, of course, every feeder believed he produced. Meanwhile, the government prohibited the slaughter of lambs before the 1st of June to increase the weight of meat available.

It was not until 1918, the last year of the war, that rationing of meat to the public was actually introduced. All fat stock was graded at the local mart and paid for by the Government. Cattle were divided into three categories and paid on a live weight basis, while sheep were estimated on a killing out weight and paid accordingly. This worked reasonably well with the auctioneer being paid the same headage fee for their work as they had been for grading and procurement.

Reaching agreement on the part to be played by auctioneers in the wartime grading and rationing scheme had not been easy and had occupied many hours of meetings, arguments and trials in 1917 with problems continuing into 1918 as a post-war report of the

Secretary of the Scottish Live Stock Auctioneers` and Salesmen`s Association reveals. This is reproduced as an appendix to this book and demonstrates the immense amount of meetings and work which was involved. As usual, the auctioneers themselves caused some of the problems and it is pointed out that some firms "did not seem to clearly understand" some aspects of the scheme and that their returns "were so inaccurate that great delay was caused in obtaining settlement".

Farming practices were changed to meet the demands of war. More wheat was to be grown as the scientists advised that one acre would feed 230 persons for a year, whereas one acre of oats would only feed 155 people. Farmers were urged to devote more attention to making silage for dairy cattle as milk and dairy products were very scarce.

The war ended and, after an interim period, in July 1920, the auctioneers got back into their rostrums and marts were humming again. Just before this time, the Government was paying up to three pounds fifteen shillings per hundredweight for beef and seven pence per pound for mutton, but prices fell rapidly, halving in five years as a high wartime-led market collapsed and an agricultural recession set in. This recession was across the board and included a massive collapse in the prices of breeding and pedigree stock, and thus of auctioneers commission. This recession was to last until 1938 when the shadows of war, again, brought Government support for farming.

CHAPTER VIII
DAIRY CATTLE IN COUNTRY AND TOWN

Dairy Shorthorns were the traditional milk cows of England, but in the period coinciding with the development of the auction marts, the Dutch black and whites steadily increased their numbers. In the 1870s there had been regular large importations. In some years as many as 70,000, of these milky, if rather ungainly looking, animals were brought in from Holland, many to Hull, across the North Sea. However, because of disease fears, the trade was restricted to slaughter only cattle between 1877 and 1889 when a short resumption in the import of dairy cattle lasted only until 1892 when all further traffic was prohibited.

The foundations of the British Friesian had been established but, inevitably numbers fell away until 1909 when a breed society was formed and further selected imports were permitted and encouraged. This dual-purpose breed expanded dramatically until the 1980s when it became intermingled with the Canadian Holstein with a resultant loss to farmers, feeders and auctioneers of the really useful dual-purpose bred black and white steer.

The Friesian breed society was founded largely at the instigation of an auctioneer, William J. Clark of Alport, Bakewell, Derbyshire who was, for a short period its first secretary. Clark was certainly a character as well as being a gambler and a sportsman. He was involved with the establishment of Derbyshire Gritstone Sheep, British Trotting Horse and Blue Albion Cattle stud and herd books. He worked for his father`s auctioneering and commission agency firm of William Clark & Co. and published the Livestock Breeders` Sale Register which put buyers and sellers in direct touch with each other at a charge of one shilling per animal.

The second and extremely influential secretary of the Friesian Society was Mr George Hobson whose brother, Harry, was the very successful pedigree cattle auctioneer. Harry Hobson was initially with John Thornton & Co. but later went his own way. By the 1970s Hobsons had three partners, David Cleveland, Robert Wheaton and Christopher Norton, who were Society Auctioneers for the British Friesian Society and for many other breeds as well as the Smithfield Show in London.

Telegrams: "SHORTHORN, LONDON." Telephone: 1020 MAYFAIR.

JOHN THORNTON & Co.

FRANK P. MATTHEWS, LANGLEY HOBBS, F. NEVILL MATTHEWS, CECIL R. MATTHEWS, HARRY HOBSON.

27 Cavendish Square,
London, W.1, England.

AUCTIONEERS, VALUERS, AGENTS AND EXPORTERS OF

PURE-BRED and MILK RECORDED CATTLE, HORSES, PIGS, SHEEP, and all other descriptions of LIVE STOCK.

ESTATE DEPARTMENT:

SALES OF ESTATES, VALUATIONS, &c., UNDERTAKEN

IN ASSOCIATION WITH

FRANKLIN & JONES, F.S.I., FREWIN COURT, OXFORD, AND W. G. MILLAR, F.A.I.

ALFRED MANSELL & CO.,

AUCTIONEERS, VALUERS, LAND AND ESTATE AGENTS,

LIVE STOCK and SHIPPING AGENTS,

EXPORTERS OF PEDIGREE LIVE STOCK TO ALL PARTS OF THE WORLD.

LIVE STOCK EXPORTERS. Foreign and Colonial Buyers supplied with any Breed of Pedigree Stock for Breeding and Exhibition purposes.

SHIPPING AND INSURANCE of Live Stock undertaken to any part of the world at lowest Market Rates. Agents to all the Leading Steam Ship Companies.

HOME BREEDERS AND FEEDERS supplied with all classes of Pedigree and non-Pedigree Stock.

Sales and Valuations undertaken for all purposes.

Enquiries invited to **Head Office, College Hill, Shrewsbury.**

Telephone: 21. Telegrams and Cables: "Pedigree, Shrewsbury."

EXPORTACION DE ANIMALES FINOS

TORROMÉ, SONS & CO.

123 CANNON STREET, LONDON.

La casa Argentina más antigua en Londres, establecida en 1860, y los *primeros exportadores* de ganado fino.

Se encargan de la compra y embarque de toda clase de ganado, aves, etc.

Exporters of pedigree stock of all kinds to South Africa, Argentina, Brazil, Uruguay, Spain, etc.

Inquiries invited.

HARRY HOBSON & CO.

Auctioneers, Valuers and Shippers of Pure-bred Livestock

Sales by Auction of all classes of pure-bred stock are conducted on behalf of Breed Societies and Breeders. Collective sales are held in various parts of the country

Pure-bred stock of all descriptions selected, purchased, insured and shipped to any part of the world

FROM JANUARY 1st, 1950 TO JUNE 30th, 1951 OUR EXPORTATIONS TOTALLED MORE THAN 2,000 HEAD OF PURE-BRED STOCK

Victoria House, Southampton Row
London, W.C.1

Telephone: HOLborn 3964 (3 lines)

Telegrams and Cables: "Pedistok, Westcent, London"

Thornton`s was latterly owned by the Matthews family. In due course, Christopher Norton and Tom Brooksbank broke away from Hobson`s and established their own business. Hobson`s and Thornton`s are long gone, Norton and Brooksbank continues to this day.

George Hobson got his first job, with the Royal Agricultural Society, on condition that he learned shorthand. This he did by sitting in the church on Sunday and taking down the readings and sermon verbatim. His brother Harry was equally assiduous and proved to be immensely successful as an auctioneer. Crewe, Slough, Reading, York and Lancaster were the favoured centres for pedigree Friesians in England, with Paisley being the Scottish centre. Lancaster became, by the 1960`s the venue for the principal heifer sale. Years before, in his first appearance at the Lancaster sales, Harry Hobson had made quite an impact; speeding up the bidding so that "an incautious nod or wink might cost as much as three to five guineas rather than the former five shillings". When selling at Lancaster one day, John Parker of Bailrig made a big thing of prodding his cow round the ring and loudly proclaiming its hidden attractions. Hobson, slightly irritated, handed him down the gavel and suggested that he had better get on with selling it himself. "Nay, nay" was the reply "I`ll carry t`gavel Mr Hobson but thee get on with thy bids, we`re doin aw reet"[20]. Hobson had first sold at a major sale, at Crewe, in 1920 when Frank Matthews of Thornton`s handed him the famous sandglass. The late John Thornton and other early pedigree auctioneers had sold with the sandglass, something that died out by the early 1930s. Like a large egg-timer, the auctioneer would turn the glass as bidding ended and the fall of the last grain, probably only perceptible to the auctioneer, equated to the fall of the hammer.

Ayrshire cattle met increasing demand throughout the period with their pedigree auctioneers, Thimbleby and Shorland of Reading holding many sales in Yorkshire, despite which many Ayrshire breeders remained loyal to Manley`s of Crewe.

The dairy trade at auction marts throughout the late nineteenth and much of the twentieth century depended not on the pedigree cattle sales of Thornton and Hobson but on the traffic of cows and heifers from their breeders to the town cow-keepers and dairymen and, later, to the many dairy farmers who maintained "flying herds" utilising bought in animals rather than replacements bred by themselves.

---

20 Gordon Mingay, British Friesians an Epic of Progress, 1982. Quoted here and in the previous chapter.

The expansion of towns and cities at the expense of the country people was the feature of British life in the 19th century. In 1851 the urban population had got to the stage that it just exceeded that of the countryside and by 1881 there were more than double the number of town and city residents as there were country-dwellers. The demand for liquid milk rose proportionately; strangely, relatively little milk was consumed per head by the countryman as opposed to the town dweller; maybe he knew what was in it!

Milk for daily consumption in much of the nineteenth century was practically all supplied by town cow-keepers with their advantages of closeness to the consumer in terms of freshness and distribution. In practice, there was a great temptation to adulterate the milk and, in due course hygiene and health proved to be a great issue.

As an example, in 1860, cow-keepers in Hull numbered 230 with an average of seven cows each[21]. Their cattle were generally kept in a shed in the back yard of a house, perhaps at the end of a terrace or on the corner of a street, with a poor water supply and no grazing land adjacent. Without refrigeration, the milk went sour very quickly so in Summer, two, and sometimes three, selling rounds took place each day after milking. A few pigs were often kept and fed on any surplus milk.

Liverpool was home to many cow-keepers. The Liverpool and District Cow-Keepers Association was established in 1865 and finished in 1975 when the last town dairy closed. Liverpool City Corporation was particularly strong on dairy hygiene issues and was one of the first cities to issue licences to operate dairy premises. In 1878 there were 452 registered premises and this rose, by the turn of the century to about 900 registered cow-keepers with about 4,000 cows between them.

From the Liverpool Mercury, December 1882:

"The first Christmas show of dairy cattle under the auspices of the Liverpool and District Cowkeepers' Association was held at Thomas Carr's Auction Yard in Breck Road, Liverpool. The interest taken in the show was manifested by the fact that there were 107 entries. The animals exhibited were a credit to the cowkeepers of Liverpool and the district, it being the opinion of most

21 East Yorkshire Local History Society. 7 cows is a very small number but fairly typical; however some of the great Glasgow city dairies milked hundreds.

competent judges that a better class of fat cattle and dairy cows had seldom been exhibited. There was a large attendance of cowkeepers, farmers and dealers from various parts of the country.

After the show, the judges, the committee and other gentlemen interested in the exhibition sat down to an excellent dinner, which was provided at Seaman's Hotel, Breck Road, Mr Robert Brownlow presiding. In the afternoon a sale of some of the prize cattle took place at Mr Carr's yard, that gentleman acting as auctioneer. There was a large attendance and good prices were obtained".

By 1949, there were only 129 cow-keepers and few of these retailed their own milk but sold it to the large dairy companies and bought it back, pasteurised and bottled. "Railway milk", brought into the city by train to these large corporate dairy companies, was the main competition and, latterly, the cow-keepers had invited inspection of their premises by the public to prove the freshness and cleanliness of their product.

The railway solved the problem of providing for seaside and resort towns, such as Scarborough, where the large seasonal variations were met by a few larger dairy farms located close to a convenient railway station. They would pay a large rent to be able to provide fresh milk, butter and vegetables to consumers, much of it sent by fast passenger train.

The heavier black and white cow was popular with the town cow keeper, who milked them for a year and then they were worth a good price to sell to the butcher. These herds were good supporters of many marts as they frequently bought a new calved cow and sold a fat one. It was to cater for them that cows and heifers were only saleable when new calved, whereas now farmers will bid for cattle at any stage in their lactation with the pregnancy pre-diagnosed.

Liverpool had some interesting exceptions to the general policy of the cow-keepers which was to buy a new calved cow or heifer, milk it dry and then sell it to the butcher. Many Dales farmers from areas such as Sedbergh, Garsdale and Dent[22] were tempted to move to Liverpool in order to better themselves as cow-keepers. They retained family farming connections in their home districts. Using mainly dual-purpose Dairy Shorthorns, the in-milk cattle would be brought to the city by rail and, when their milk supply dwindled, returned to

22 Joyce Scobie, Cowkeepers from the Dales, Sedbergh, 2008.

the family farm to be put in-calf again or fattened as appropriate.

The First World War, with the men mainly away in the armed services, and the arrival of electric refrigeration[23] gradually reduced the numbers of the town milkers, much to the delight of the Medical Officer of Health who had found in 1880 that roughly a fifth of all milk samples taken were adulterated, usually with water. The producing farmer-retailer greatly improved the quality and presentation of the milk and, after the Second World War, he was replaced by the multiple retailer[24] of pasteurised milk under the auspices of the National Milk Marketing Board which was, mistakenly, abolished by Mrs Thatcher`s Conservative government as a result of EU pressure.

A city dairy, Batty & Sons, Aigburth Road, Liverpool

In 1937 the eradication of Tuberculosis from both dairy and beef herds became a priority and after the war progress moved very rapidly spreading Tuberculin Attested cattle from the South West of Scotland into Yorkshire, Cheshire and Lancashire. The Scottish Milk board had given an extra one penny per gallon to attested herds, which by 1940 numbered 16,000. Finally the U.K. was declared T.B. free in

23 One of the first refrigerators in Liverpool was installed by a cow-keeper, George Batty of Aigburth Road; other pioneering introductions by the Liverpool cow-keepers included the early use of cattle drinking bowls and the first use of the electric milk float.

24 Express Dairies Distribution was opened by Robert Barham in 1864 with all the milk supply being brought about 20 miles by train in to London.

1960. It was reckoned that, before eradication, about one quarter of the national dairy herd was infected with the disease. Transmission to humans was often fatal and caused very large loss of life over many years.

How sad and stupid that half a century later TB has been allowed to re-infect a large part of England's cattle causing many problems for farmers and, of course their auction marts, as well as the veterinary and health services. Over 35,000 infected cattle were culled in 2015. Official inertia and a fear of the badger enthusiasts has exacerbated the problem.

Bovine tuberculosis having been eradicated in the sixties, brucellosis was defeated in the seventies. With all milk now conveyed in tankers, the traditional twelve-gallon railway pattern milk can celebrated its centenary by becoming a collectible antique, soon to be followed into oblivion by the roundsman milkman as the supermarkets took over.

As far as auctioneering was concerned, dairy cattle sales were an important part of the mart business and, by 1890 over fifty auction marts in Cheshire, Yorkshire and Lancashire sold milk cows every week. Advertisements showed varying numbers of between six and sixty, the exception was in Preston, where every alternate Monday was held the largest dairy cattle sale in England. One report mentioned that 279 were forward at the Farmer's Auction Mart, Preston where Mr Hothersall was auctioneer. He later acquired the business. About three quarters of the milk cattle sold were Shorthorns but the Ayrshire and Friesian breeds continued to grow in popularity. Jerseys were favoured by many home farms of the gentry who supported sales in York.

By the 1930s advertisements promised that Hellifield would have, weekly, 350 calved and in-calf cows and heifers, Skipton would have 300, Preston 300, Otley 250, Gisburn 500 and Lancaster 150, as well as there being many smaller sales. Henry Manley from Crewe also held frequent sales of dairy cattle at Yorkshire marts.

In Yorkshire and Lancashire there were many small farms, which carried a small flying herd of dairy cows bought new calved at the local mart and, when dry, sold for slaughter, again at the local mart. This system that brought constant business to the auctioneers who were supplied by many cow dealers who, every week, brought Shorthorns from the marts in Cumbria and Ayrshires from the marts of South

West Scotland, where my own marts at Annan and Dumfries could sell up to six hundred calved cows and heifers at a weekly sale. The Ayrshire/Shorthorn cross was a popular dairy cow. Later Friesians and Holsteins took over from the Shorthorns and, in turn, superseded the Ayrshires.

The animal health schemes to eradicate Tuberculosis and, later, Brucellosis forced the larger flying herds to change to breeding their own replacements in the late 1960s by which time commercial new calved dairy cattle were making around £200 each.

From the end of the war, from 1946 onwards, the auctioneers invested huge sums of money and effort into providing facilities for the eradication of Tuberculosis from the cattle of Britain. New, segregated mart buildings and new sales sprang up all over the country and after twenty years, Bovine T.B. was only a bad memory; but this was not enough as Brucellosis was the next disease which had to be routed with the wholehearted support of the auctioneers but to their detriment as more and more farmers established closed herds rather than buying their replacements through the auction marts.

In 1984 milk quotas were introduced to reduce European Union milk production by about 10% and to bring an end to the "milk lakes" and "butter mountains" which had built up with the intervention buying process. This reduced the value of dairy cattle with a consequent loss of commission for the following couple of years.

The Lister family at the Bridge End Auction Mart at Otley had become, in 1951, the first cattle auction mart in the country to confine its entries to T.B. Accredited cattle and the mart there continued to flourish for fifty years until its closure in 2000. On many sale days over four hundred dairy cattle were sold in Otley mart and special sales of pedigree cattle were held for Friesian, Ayrshire and Jersey cattle breeders. The Pennington brothers held an annual sale there of Friesian cattle when over 700 cattle were squeezed into the mart and Tom Bell from Scotland also provided Ayrshire calved heifers for two annual single owner sales. Among these special events, Addison`s held an Autumn sale of 120 Ayrshire calved and near calving heifers for John Metcalf of Bolam Grange and Haslingden Mart would hold sales for Winter Brothers of Appleby who were one of the largest cow dealers in the north. Manley`s of Crewe also conducted many such single-owner special sales.

At Holmfirth, the auction company proved reluctant to establish T.B.

Attested sales and so the local farmers set up an opposition business, The Holmfirth Attested Auction Mart, which continues to this day.

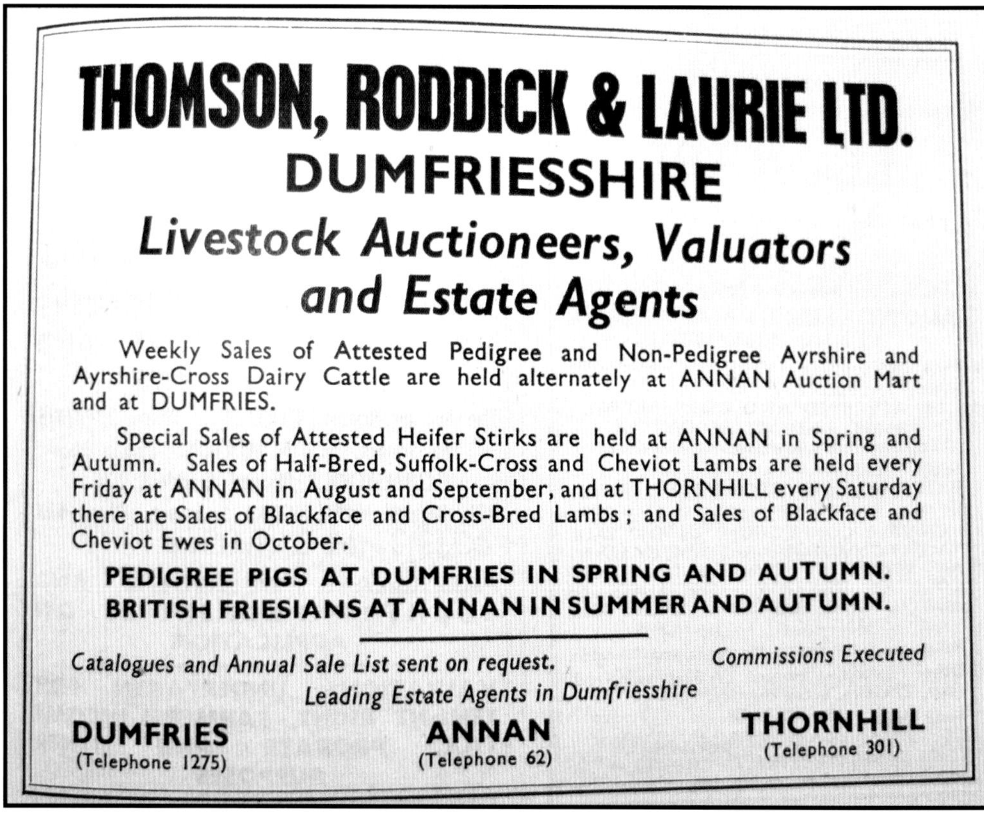

The author's advertisement, 1951

At that time, one of the Otley directors was Alan Harrison of Earby whose brother, Dick, bought cattle from me every week at Dumfries. Their father, Richard, was always remembered as one of the first dealers ever to give me a tip. One shilling for helping to drive his cattle a mile to the railway station when I was about ten years old.

To this day, Messrs. Norton and Brooksbank, now of Cirencester, hold sales of mainly pedigree dairy cattle on farms, or jointly with marts in the three counties. In 2016 the Cheshire family of Hassall Brothers started to hold weekly sales at Barber's Mart in Market Drayton, where considerable numbers were sold, then in 2018 joint monthly sales with NW Auctions at Lancaster were announced. At the first sale in March, 44 dairy cattle were forward and sold well up to £3,000 and averaged £1,764. In their first two years at Market Drayton they sold 3,500 dairy cattle, which exceeded their expectations.

In the 1960s and 1970s dairy cattle sales were held every week at:

Bentham
Blackburn
Beeston Castle
Chelford
Clitheroe
Chester
Congleton
Crewe
Doncaster
Gisburn
Haslingden
Hawes
Holmfirth
Lancaster
Leyburn
Northallerton
Otley
Preston
Rawtenstall & Bacup
Sedbergh
Skipton
Wigan
York

After 2010, only the following marts continued to hold regular sales, some only fortnightly or even monthly:

Beeston Castle
Bentham
Chelford
Gisburn
Lancaster
Otley
Skipton
York

CHAPTER IX
IRELAND AND IRISH CATTLE

In the late 1870`s hostility between the, usually Roman Catholic, smallholders and tenant farmers of Ireland and their, generally Church of Ireland Protestant, landlords developed to the stage that the former were persuaded by radical politicians to form a Land League with the general aim of pushing out the "ascendancy" landowners and appropriating their land for the people. Over succeeding years, acts of parliament were passed providing government money to buy estates and arrange cheap loans to tenants to enable them to acquire the land. By 1913 eight million acres had been purchased in this way for an average price around £10 per acre. One of the tactics adopted by the League, particularly in 1879 and again circa 1890 and 1891, were attempts to drive the landlords out by disrupting their home farming enterprises. Stopping the sale of their cattle was a favourite ploy and as this report from The Herald of 6th Sept., 1890 shows, it extended to England with Salford Cattle Market particularly featuring.

Boycotting at Salford

Since the conviction at the Liverpool Assizes, some time ago, for boycotting in the Salford Cattle Market, there has been no interference with dealers, but this week an attempt at boycotting was frustrated by the prompt action of the police. During the early part of the day considerable excitement prevailed in the centre of the market, owing to the presence at the stand of Messrs Lambert & Son, cattle salesmen, of some heifers which had come from a boycotted farm in Kildare. The cattle had been watched and followed by Land League agents from the farm, until they reached the Salford Market. As soon as they were placed in the pens several men, said to be Land Leaguers, took up positions near them, and for some time walked about the pens. The Irish police had been informed of the intended movements of the League men; a couple of detectives had accompanied the cattle from the port of embarkation, and they had kept a careful watch upon the movements of the League agents. At the Salford Market two other Irish detectives and several Salford detectives and members of the police force joined the officers who had come from Ireland, and prevented any interference whatsoever with Messrs Lambert, and the cattle were sold, whereupon the League agents left the market. Messrs Lambert expressed their determination not to be interfered with in their business and threatened to have the men locked up. Some difficulty was experienced in getting the animals removed from the railway waggons, owing to the refusal of drovers to take them, and the leg of one of the animals was injured. Some suspicion is entertained that the injury was purposely inflicted.

Inspection of Irish cattle landing at Birkenhead

There was a long history of Irish cattle being brought over and it is recorded that some 60,000 Irish cattle had been imported in 1663. However, the punitive and partisan Irish Cattle Bill of 1666, which was promoted by those whom Samuel Pepys called "the western gentlemen", landed gentry in the west, north and Wales who, thinking that the value of their own cattle would be increased, put a stop to this trade for a long period. Indeed, cattle imports from Ireland were only re-commenced late in the 18th century; and it was only after the terrible Irish famine of the late 1840`s, when some million and a half people perished and another million or so were forced to emigrate, that the trade hugely expanded. As the cottagers and small farmers left the land, either through choice or compulsion, farm sizes rapidly increased. Holdings of less than five acres made up almost 50% of the total of Irish farmland in 1831 but by 1851 this had reduced to only 15%. Non-dairy cattle numbers probably doubled between 1841 and

1851 and, in 1849, Irish cattle exports to mainland Britain numbered some 211,000 head[25]. Until well into the century the majority of the imported cattle were by the English Longhorn crossed with the native breeds but by the 1850`s the Shorthorn breed dominated.

By 1854, it was reckoned that relatively large "lease-holding farmers" were occupying 75% of Irish land[26] and the land given over to pasturage increased by nearly 2,000,000 acres between 1851 and 1901 as arable ground similarly decreased. In 1851, 400,000 acres were under wheat and by 1901 this had reduced to less than 48,000 acres. Aside from a depression in the period 1859 to 1864, Irish farming remained fairly prosperous throughout the second half of the 19th century. Apart from the north-east of the island (Ulster) and the Dublin area, Irish industrial and commercial output was small and there was an almost total dependence right through until the mid-20th century on the success of agriculture. The lynch-pin of this was the all-important export trade of live animals to England and Scotland where the trade provided business for the marts and essential supplies for the cattle feeders who, in turn again, filled the marts with fat cattle. This was particularly important to the Yorkshire auctioneers.

George Strachan, who managed the Northumberland firm of Maughan's auctioneers, writes in his autobiography "it took a specialist auctioneer to deal with and sell Irish Cattle, as the Irish cattle trade was quite different from any other auction sale." My father, who had sold many Irish cattle, told me that those from County Sligo were the best doers but I only fed Irish cattle when I took pity on some poor old friend who still had not managed to sell a pen late in the day when all the farmers had gone home. He would then go to the station waiting room to get the 4 o`clock in the morning boat train for Stranraer and I would lock up the mart.

In these days there were few auction marts in Ireland, the Irish remaining attached to their fairs. There was a considerable mart near Enniskillen and a very large one in Dublin where Ganley`s did an immense business with many cattle shipped to Holyhead and Birkenhead. Birkenhead, on the Mersey, had large lairages at Woodside where the cattle were all checked by the Ministry veterinary surgeons. From the port there would be a railway or lorry journey to the fields near the marts where the animals rested for a day or two before being walked in the early morning to the market where they

25 In the same year , cattle imports to England from Continental Europe were about 40,000.
26 C. Woodham-Smith, The Great Hunger, 1962 & R.F. Foster, Modern Ireland, 1988.

might either be sold by private bargain or by auction.[27]

Each week in the autumn over four thousand cattle would be shipped from Dublin to the Mersey to be carried onwards on the railway, many to the marts of Yorkshire where they would be sold to the feeding farms of the East Riding and Lincolnshire, in due course producing large entries of top quality cattle for the auctioneers` prime stock sales. Some weeks in York, some two thousand Irish cattle would stand for sale in pens around the city walls. The best heifers went to Lincolnshire whilst the Yorkshire buyers tended to prefer bullocks.

Unlike Northumberland, where very large numbers of Irish cattle were sold by auction, Lancashire and Yorkshire had only two markets advertising and reporting regular sales of freshly landed cattle. These were York and Northallerton, complemented by the County Durham mart of Darlington, which supplied many Yorkshire graziers. On the east side of the Pennines many Lancashire and Cheshire farmers went to the landing lairage at Birkenhead to buy direct from the importers, where arrivals in one week in September 1935 were reported as 2,338 cattle and 1,811 lambs from Ireland.

In Yorkshire and Lancashire, most of the Irish cattle dealers did not use the local auctioneer. They traded in the markets that were owned by the city and town councils and who would charge them a toll for the use of a pen, which was all that was needed. They had a never-ending patter which enabled them to sell their cattle direct to the farmer without paying commission to the auctioneer. The auctioneering firms co-existed with this arrangement, knowing that they would in turn see the cattle in their prime markets. In York, even after the Stephenson and English consortium built their fine new mart, they continued to let the Irish dealers have pens to sell their cattle privately as York Corporation had done from time immemorial.

Only very occasionally would the Irish dealer come to the auctioneer to tell him he was in a generous mood and, as requested by the buyer, he was going to book the sale of so many cattle through the mart. This unbounded generosity was often accepted very reluctantly by the auctioneer, who knew very well that that particular buyer could not pay until he sold the cattle fat with him in several months time.

A number of prime cattle from Ireland were sold every week at Salford

27 At some local authority markets the pens for the Irish cattle often had no front gate so that drovers or their dogs had to be constantly on duty to prevent escapes. I never heard a good reason given for this curious design.

Market and at Stanley Market, Liverpool where also a considerable number of Irish store cattle were sold for further keep.

In 1923 regulations were introduced stating that cattle from The Irish Free State had to be isolated for six days and that auction marts had to obtain a licence to sell them. The days when Irish cattle dealers could rush cattle from one market to another before selling were consigned to history. A dealer once told me of shipping two hundred and twenty cattle to Birkenhead - five days and four marts later he had them all sold expect one that "had died of exhaustion".

In the years after the Great War, imports from Ireland to the United Kingdom averaged some 400,000 fat cattle and 500,000 store cattle. Subsequently quoted numbers applied only to Eire as Ulster was looked upon as part of the United Kingdom.

In 1913, 655,000 sheep had come over from Ireland and by 1914 835,000 cattle were crossing the Irish Sea annually and these high numbers were maintained until 1965 when the Government of Eire encouraged the slaughter of many cattle to provide an additional source of employment and added value for the economy.

The Irish store cattle trade died over the succeeding twenty years as many British arable farmers replaced the essential fertility which had been so long provided by cattle dung with artificial fertiliser. The marts serving the arable farms of the East Riding and the East of England in general were particularly hard hit.

T.B. testing found many Irish cattle to be infected, which reduced the demand in Britain from 1960 onwards and, at the same time, the Irish Government was, as previously mentioned, encouraging the provision of abattoirs in Eire to concentrate on the export of meat.

Many meat plants in Ireland were acquired by the controversial Larry Goodman who also bought nine plants in Britain, employing 1,600 workers by 1985. He had a considerable influence on trade and specialised in supplying supermarkets with stock direct from the farm. Later the Goodman group was to become Europe's largest beef processor with their empire including slaughter plants in Poland. Such developments, along with the increasing dominance of supermarket buyers, did not benefit the auction marts.

When writing of Ireland, I am reminded that, in 1888 Patrick Leonard produced an excellent book of tables for the United Irish Cattle Traders Association at Dublin. The tables include conversion of liveweight prices to deadweight prices per pound, per stone and per hundredweight then it continues with conversions for waggon loads of up to five tons, sheep conversion tables and allowance tables for travel losses where the loss of deadweight meat is not fully reflected from the liveweight loss, which includes elements not affecting the weight of the meat, e.g. excrement. He concludes by saying that "knowing the weight only partially overcomes one of the many difficulties that attend the business of the cattle trade".[28]

For some years, during the drive for Home Rule and independence, it was not easy being an auctioneer in Ireland, as the I.R.A. burned down the mansion houses and boycotted the Protestant land agents, auctioneers and cattle dealers who had served the estates. Even after 1921 when an Irish Free State was formed, a few extremist I.R.A. continued a civil war which caused ruin for many Anglo-Irish landowners and their supporters, until the new Garda police force gained control in about 1926.

At sales of the crop and estate effects, the auctioneer would be told by the local I.R.A. thugs that he must only take bids from the local smallholders; if he refused he was boycotted and the crops he had sold were burned. Several auctioneers fled to England to escape being murdered and had to leave their houses and furniture to be destroyed or stolen. Protestant dealers who had moved the estate cattle were also boycotted and forced to move to England where the government provided a house but no compensation for loss of their business.

From 1919 to 1926 both London and Dublin provided some compensation for damage during the Civil War. The estate factor would report to the owner, often already resident in England, that all was destroyed by fire, lodge a claim for compensation then be off across the Irish Sea where often he found it difficult to find a job.

Among the auctioneers ruined was John Holmes of County Limerick, who was pressured by I.R.A. notices warning people to do no business with this ex "British spy". He then held a sale with a large crowd but not a single person made a bid. Among the small tools a few women and children called out "one penny, one penny" and Holmes said, 'it's yours" as he brought his stick down on the boards of the old cart he was using as a rostrum. Needless to say he got no more sales. Michael Quirke, an auctioneer and land valuer in County Tipperary was boycotted and burned out for having done most of his business with the land owners. In truth, they were the only people

28 The 80 page book cost 3 shillings or 15 new pence and his price range was from £1 to £2 per live hundredweight

that he could earn any money from; however, he obtained £70 of government compensation and left the district alive. A number of Protestants went to the North of England where they got a living as drovers or salesmen for Irish cattle dealers.

Advertisements and sales reports of Irish cattle in Yorkshire, Cheshire and Lancashire are rare compared to the remainder of England, where the auction companies were more actively involved in the sale process and ensured that they were publicised with results recorded in the local newspaper.

The following excerpt from the author's book on the marts of the far north of England shows the scale of the sales and landings recorded in a newspaper of 16th September 1935.

Alnwick	2,000	Acklington	800
Bishop Auckland	150	Scots Gap	200
Berwick	200	Tow Law	300
Gateshead	900	Silloth Dock	700
Hexham	500	Glasgow Dock	4,200
Morpeth	300		

To improve marketing, the government in 1927 made it compulsory to weigh all fat cattle before sale by auction. The Irish importers claimed their cattle were store not fat, so many disputes took place. However auctioneers generally welcomed the Act. Statistics show that in the years up to 1970, imports of live store cattle from the Irish Republic fell from 620,000, in 1967 when they contributed £40,000,000 to the Irish economy, to 243,000.

CHAPTER X
A LITTLE ABOUT SHEEP

By 1867 and 1868 the prices of wool and sheep had collapsed to such an extent that in 1869 many wool fairs reported that no prices were quoted. It is suggested that one of the causes of this total collapse was the end of the American Civil War which resulted in a glut of imports to Britain.

Such a collapse in values did not occur again until 1932, when many farms in England could not find a tenant and others only remained with help from their local auction mart. Charles Mercer from Burscough would take loads of Cheviot lambs, out of Scotland, to Lancashire that had only cost ten shillings (50p) each, and still he struggled to sell them.

At the sheep fairs of the mid to late 19th century, as the afternoon arrived but no buyers came to give a worthwhile price, the local auctioneer would be called upon to try his skills. This often proved successful and resulted in an increasing number of transactions with the outcome that, each year, more were sold by auction. These fairs became important and enjoyable annual events in the auction calendar; held, as they were, in temporary pens in the market towns or in the isolated villages of the Dales and Moors of Lancashire and Yorkshire[29].

In the East Riding, Sir Tatton Sykes of Sledmere (1772 – 1863) was, as well as being a noted horse owner and breeder, a major sheep enthusiast and an early adopter of the auction system. His annual sheep sale drew buyers from all over Britain and ran for fifty eight years, the last being held in 1861.

The North of England was historically noted for the annual sheep sales or fairs held in the moorland districts, usually in or close to a village or public house. Serving the North Yorkshire Moors were sales such as Fadmoor and Wombleton[30] which still take place today and Ingleby, near Stokesley, where H. Richardson and Robert Farrow held two annual autumn sheep sales at the Dudley Arms; and three miles away, at Kildale, a sale was held by R. and W. Hedley for the Kildale

29 There were many of these rural seasonal sales and often there is little information about them extant. There has been no attempt to cover them comprehensively in this book or its appendices but some are highlighted by way of example.

30 At the end of season Wombleton Sale, 2018, Cundalls sold over 6,000 sheep.

Sheep Society. From the 1880s, three large autumn sales were held, by Richardson and Smith of Whitby, at Glaisdale and Castleton. The first of the Glaisdale sales, in early September would have comprised about 3,000 lambs. In the Dales country, sales included Reeth, Kilnsey and Malham. At Malham, the stock were penned on a site where the monks of Bolton Priory would have gathered their sheep some eight centuries before. The pens for each flock would be on the green in the same place as they had been since time immemorial. It was auctioneer, T.H. Taylor from Skipton, who, in 1920, transformed Malham Sheep Fair into Malham Annual Auction Sale of Sheep where he gathered a good ring of supporters, who would watch at twelve noon as the seller's names were drawn out of a hat to decide whose stock should be sold first and then, in the autumn air, the sheep would move quickly through, being sold to bidders who would come back year after year to buy their stock.

Sheep Sale at Kilnsey
(North Yorkshire County Archives)

Glaisdale, Kenneth Ernest Halley, auctioneer

Glaisdale, 1987, Robert Smith selling, Ian Halley clerking, Richard Smith looking on behind auctioneer

Sheep Sale at Fadmoor, 2005
(Photo, Colin Grice)

These seasonal sales are now much reduced in number, but still remembered fondly. Their decline started with motor lorries bringing centralisation, followed by decreasing numbers as sheep stocks fell and the continuing sheep farmers looked for more productive ewes. Some feeding their lambs and sending them to the prime ring or direct to the abattoir. Afforestation has played a large part in reducing sheep in some areas. Not only the marts but the great mills that employed so many people turning the fleeces of the sheep into cloth, have gone.

Sheep still play an important part in the calendar of most marts and many weekly prime sales and autumn seasonal store and breeding sales are still held, which I will review briefly going from east to west.

Starting in the east, Lancaster sell store lambs every fortnight with Mules, Mashams and Continental breeding sheep in between. They also have large sheep sales at their new mart at M6 Junction 36 in Cumbria, which replaced Kendal and provides competition for some of the relatively nearby Lancashire and Yorkshire marts.

Clitheroe sells store lambs weekly on Monday, with lots of special sales including those for the Lonk Sheep Association, the Derbyshire Gritstone Society and the North Country Cheviot Society. The "Old Fair" sales at the end of October still attract large numbers of breeding sheep with North Country Cheviot rams to 1,420 guineas, and Lonks and Gritstones both to 4,000 guineas in 2016.

Bentham and Gisburn, where Richard Turner's family preside, along with Stephen Dennis, are always busy. Bentham holds large sales on several days of the week including cull ewes drawn from a wide area and with over a thousand rams featuring on one Saturday. Gisburn sells every Saturday and has a seasonal "twilight sale" of rams, which reminds me of my early years with the flickering light of the oil lamps, but I am sure bright electric lights now show the bidders clearly. Still in Lancashire, Brockholes, near Preston, holds smaller weekly sales of store lambs.

Hawes continues to be the most famous store sheep mart with 70,000 finding new owners on ten days of sales each autumn with Raymond Lund in command. On two days in September, over 30,000 Mule ewe lambs are sold, a fortnight later 12,000 Mule shearlings and ewe lambs come to market. Over a thousand Bluefaced Leicester rams and ram lambs attract a multitude of buyers with eighteen hours selling running into the early hours of the morning after the day before and with many exciting prices being achieved right up until the end. A fortnight after this sale some 800 Swaledale rams take the stage over two days with recent years seeing bidding to a top of £92,000 from the Risdon flock, followed at £86,000 and a wonderful average of £2,226.

Malham sheep sale, 1969
(North Yorkshire County Archives)

Twenty miles away at Leyburn, managed by Stephen Walker, Friday sales are held of Swaledales, Mules and other breeds, with an annual show and sale of Teeswater sheep and concluding with the Middleham Moor Fair sale. Travelling east for another twenty miles, Northallerton mart holds sheep sales on Wednesdays, then only ten miles south is Thirsk, where special sales are held on Saturdays, including the North Yorkshire Fair in mid-September. The Northallerton Auction Co., also sold sheep at Bedale mart until its closure. Reeth was the site, until 1992, for two popular annual sheep sales which had been conducted for many years by J.R. Addison and Son.

The sheep country of the North Yorkshire Moors was served by a ring of marts - Guisborough; Ruswarp, by the Moors Railway, near Whitby, which holds many special sales; Helmsley and Pickering. Only Ruswarp remains, successfully run by Richardson and Smith of Whitby. John Deacon held two annual sales at Ravensworth until 1930.

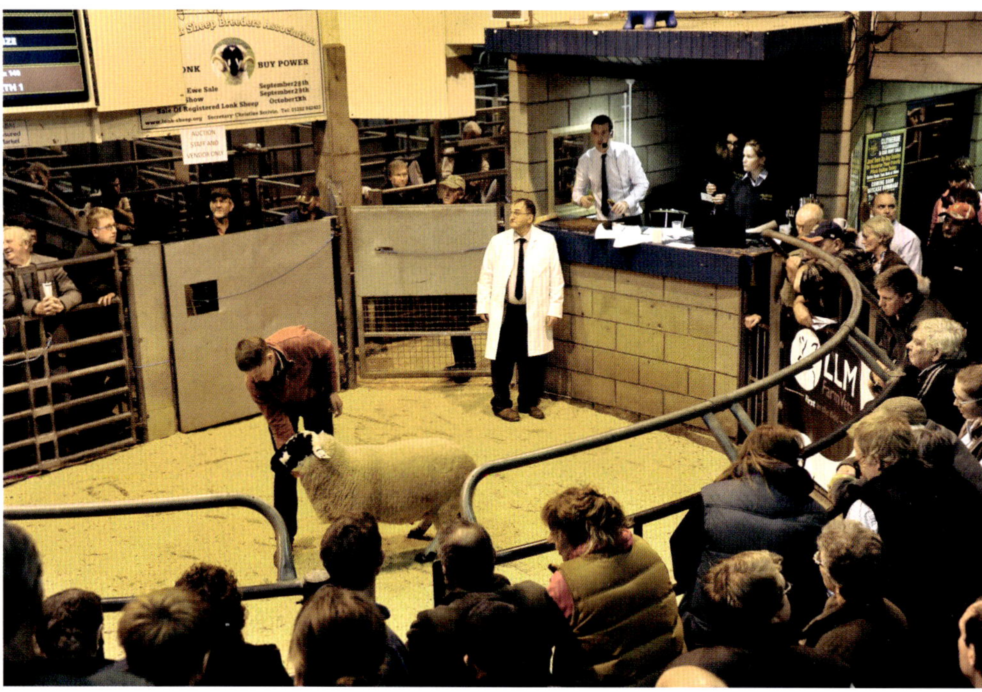

Clitheroe, 2018 Derbyshire Gritstone, 4,500gns, Joe Worthington in the ring, Jeremy Greenhalgh on the rostrum

(Photo, Farmers Guardian)

Malton hosts some large sales with some still held traditionally in temporary pens in the town streets, Malton late fair saw 7,000, a record entry last year. Cundalls of Malton hold seasonal sales at

Fadmoor and 4,000 are sold at Wombleton in the North York Moors, where Keith Warters is auctioneer and Tot Collier has been droving the sheep for over seventy years.

In the heart of NorthYorkshire, Skipton has a particularly fine auction centre, where Jeremy Eaton and Ted Ogden gather large sales of sheep every week in the autumn, with many special days for prize shows of several breeds. They start in mid-July with a wonderful sale of 5,000 store lambs and then keep going with their September sales including Wensleydales, Jacobs, Beltex, Charolais, Texels, Lleyns and Hampshires; the full spectrum of breeds with Mashams in early October followed by Dalesbreds, the traditional mothers of the Masham.

Moving east, Wharfdale Farmers Mart at Otley sell Mashams and other breeds and York on Thursday, Selby on Saturday and Dunswell, hold sheep sales but all sell more prime animals than store or breeding sheep.

Leyburn Mart, 1994, David Wallis selling Masham gimmer lambs, seated are Don Wrathall and Michael Peacock

Hawes 2016, Raymond Lund auctioneer

Sheep Sale at Fadmoor, 2005
(Photo, Colin Grice)

In the South of the county, Doncaster held weekly sales until closing in 1990 and sheep continue to be sold at Holmfirth with some autumn sales in conjunction with Paul Dixon of William Sykes and Son.

The great sales of Masham sheep at Masham Farmers Mart, the Wensleydales at Hellifield and Rough Fells at Sedbergh are now only a memory, as is the annual sale of over 2,000 gimmer shearlings, sold by Andrew Oliver of Hawick for John Nickels of Fridaythorpe in the East Riding, who had bought them all as ewe lambs in the Scottish Borders. Latterly, this sale was held in conjunction with Dee and Atkinson.

Masham Sheep Fair, 2013

In Cheshire, Beeston Castle holds many sales, but the county is surrounded by the great sheep marts of Wales.

All these store sheep sales depend upon a large number of experienced friends who, year after year, can be relied upon to bid rationally, buy reasonably and pay readily for a commodity with no known end price. A successful store sheep market depends on a reliable prime sheep market and we are fortunate that the sheep feeder has had the sense to maintain strong support for the live ring, unlike his compatriot who fattens cattle. The live market is extremely responsive to market forces and protects the seller from being held totally to ransom by the processors and supermarkets who seek the demise of the prime sheep mart in order to further control the supply chain to the farmer's disadvantage.

When I was a sheep auctioneer we studied the Christian calendar to find out the date of Easter, when there was a big demand for new season lamb, which the mart must satisfy. Now the sheep auctioneers must study the Muslim calendar to find out when the month-long celebration of Ramadan is, which culminates in the Feast of Eid, for which there is a huge demand for mutton, which the marts must satisfy. How circumstances change.

CHAPTER XI
A MENTION OF HORSES

Although my researches have focussed on agricultural livestock sales, it is difficult to write about Northern England and Yorkshire in particular without a brief mention of horses; and what a lot of horse auctioneers thrived in Yorkshire, Cheshire and Lancashire; and so many breeds, from the native Cleveland Bay, which pulled the mail coaches in the East Riding to the Shires sold at Preston and the Thoroughbreds at Doncaster, with foal sales, yearling sales, end of season sales ranging from holiday seasons to hunting seasons, weekly sales, monthly sales, annual sales and lots of knacker sales to complete the calendar. Regular horse auctions pre-date agricultural sales by at least a century.

Tattersall's Doncaster Horse Sales, early 1950s

In Liverpool, Mr Lucas[31] of Hotham Street was a major auctioneer of horses. He held sales on each Wednesday throughout the year with cart, van and harness horses, together with carts and vehicles. On Saturdays he held rather better sales which were for hacks, harness horses, cobs and ponies accompanied by carriages, harness and

31 His business wasn`t restricted to horses. In 1890 he was advertising a two-day sale of Channel Islands cattle at his repository.

saddlery. Horses had to be delivered at least a day prior to the sale to avoid the risk of stolen animals being brought in off the street and to give the auctioneer`s grooms the chance to spot any undisclosed faults before the animals were offered. Select horse sales were held fortnightly on a Thursday and, for these more upmarket events, the animals had to be forward six days prior.

Among other leading horse auctioneers in the region were Walker & Sons in York, Poultney`s and Hepper`s in Leeds, Boulton & Cooper in Malton, Walby in Sheffield, Hothersall in Preston, Bennett in Manchester, Armitstead in Lancaster, Manleys in Crewe and Wright in Chester. In 1900, Armitstead was promoting Cockerham Great Christmas Horse Sale as "the most successful sale in the country" and, in October 1901, Bainbridges of Warrington were advertising end of season sales of tram horses for the Lancaster and District Tramways Company and the Morecambe Tramways Company. In all, there were some sixty horse marts, or "repositories" as they were known, in this area. All have passed into history, but Stephensons of York still maintains the tradition with a monthly horse sale where the best Mountain and Moorland ponies show their popularity, now mainly used as children's riding ponies, but some making very substantial prices. Back in around 1900, useful cart horses were selling for £50 to £70 with good carriage horses from £70 to in excess of £100.

There was, of course, huge demand for horses of every type and for every use from pit ponies to coach horses and from agricultural draught horses to horses for urban transport. In regard to the latter, horse drawn omnibuses came into use in the early to mid 19th century. In 1823, John Greenwood, a toll gate keeper on the Manchester to Liverpool road, attended an estate auction and bought a cart which had been converted to have several seats and a roof and had been used to carry the domestic staff to church. Buying the horse also, he commenced the first regular omnibus service in Britain, from Pendlebury to Manchester and return for the cost of a threepenny token. By 1856, the London General Bus Company was stabling some 500 horses to draw their fifty omnibuses[32].

Incidentally, one of the best coach builders in Lancashire was Mr Gorst of Great Charlotte Street, Liverpool, who built coaches for the aristocracy. His problem was getting payment as the grandees were

32 The first omnibus in London had trotted out in 1829 and the ensuing chaos caused by buses picking up on both sides of the road was not resolved until 1867 when parliament enacted that traffic must travel on the left.

renowned for leaving their bills unpaid for even two or three years to the embarrassment of the tradesmen who supplied them.

Yorkshire is, of course, renowned for its connection with Thoroughbred horses and, as far as auctions are concerned, it is believed that the Doncaster St Leger Festival Yearling Sales are the oldest regular sale fixture in the UK.

Tattersalls, the oldest bloodstock auctioneers in the world and established in London in 1766, had been involved at Doncaster since 1828 and took over holding the yearling sales there from 1840, on the death of John Boulton who had previously conducted them. At that time, the sales were held in front of the Salutation Inn, close to the town centre, moving in 1859, for a short time, to the Horse Fair Ground. This was succeeded in 1867 by a move to purpose-built premises at Glasgow Paddocks, named for the Earl of Glasgow who had kept his horses there. Here, for almost a century, Tattersalls developed the world-famous sales with prices and averages often exceeding those of their Newmarket counterpart.

By 1955, the excitement over the visit of the Queen to the sales belied the undercurrent of uncertainty which was developing with regard to the expiry and renewal of Tattersalls` lease, the needs of the site owners, the local council, and the unwillingness of either of these parties to invest in updated or new premises. By this time it is recorded that "the sale ring was pitiful in comparison even with the modest standards ruling in the Park Paddocks at the time. The ring consisted of a single white circular post-and-rail fence located in a paddock of tufted, unkempt grass and the horses wore down a bare earth track round the inside. The rostrum was in a ramshackle wooden shed. The stand was primitive, as if it might have been obtained from one of the minor National Hunt meetings." Final closure after the 1956 Sales was attributed by Peter Willett in "The Story of Tattersalls" to "the expiry of the lease, the Corporation's requirement of the Glasgow Paddocks for civic development and the prohibitive cost of building suitable facilities at another site in the town."

Many in the racing world were philosophical about the closure and transfer of the sales to Newmarket but not so local businesses. The owner of Ye Olde Bell Hotel at Barnby Moor, bemoaning the loss of the wealthy international racing set who would attend the sales said: "Doncaster will find themselves catering for parties who bring their own lunch and buy bags of fish and chips. Anyone who thinks that the absence of the yearling sales from St Leger race week will not affect trade is a fool".

There would be no sales at Doncaster for five years. In the early 1960s after much local pressure, the trainer Willie Stephenson was persuaded to recommence these at the racecourse stables. His partner was that distinguished racing man, Ken Oliver of Andrew Oliver and Son of Hawick, the great Borders livestock auctioneers. Trading as Doncaster Bloodstock Sales, they commenced sales in 1962 and, joined by Harry Beeby in 1964, made great strides in developing a progressive and successful business with worldwide connections and an innovative and varied sales calendar. In 2007 they joined with the renowned Irish bloodstock sales company, Goffs and the merged business continues to go from strength to strength with, in 2015, twenty-five days of sales realising over thirty seven million pounds.

Preston Auction Mart, Shire horse sale,1936

The importance of horses in every sphere had been brought home back in 1918 when it was reported that over two million horses had gone to war; of these, about half a million were from North America. Horses had fought in the first battle of the Great War when, on 22nd

August 1914, a squadron of dragoon guards scattered a German unit east of Mons, taking many prisoners. Over four years later, on the very last day of the war, only twenty minutes before the armistice hour of eleven o` clock on 11th November 1918, British cavalry galloped ten miles to capture an important river bridge which opened the way to Belgium. Again they took prisoners – 171 Germans who thought that the war was already over. British horses on the Western front were looked after by some twenty specialist horse hospitals. After the war, there was an immense surplus of military horses with auctioneers conducting sales of these in marts all over the country. Demand for agricultural use was less as, during the war, 6,000 or more tractors had commenced their work on British farms. Mules were acclaimed by many military men to have been worth their weight in gold during the conflict with their stamina and hardiness eclipsing those of the horses but, unfortunately, when brought to market after the war, they proved almost unsaleable.

Rushworths wagonette, circa 1900

A chat at the horse sale

CHAPTER XII
BETWEEN THE WARS

The politicians of 1919 promised that the future would be "a land fit for heroes to live in". For farmers this heroic promise lasted only for about two years as prices fell by a quarter. Much worse was to come as by 1932 prices of farm produce had collapsed by two-thirds and many farmers were bankrupt and homeless. Our family hill farm recorded their Cheviot wether lamb prices falling from forty five shillings to fifteen shillings (75 new pence) per head.

Shorthorn and Ayrshire pedigree cattle dispersal averages fell from about £90 to £30 and the top price at Kelso Ram Sale dropped from £1,100 to £150.

The collapse from prosperity to disaster was so swift that many auction marts feared for their future. Some called for commission rates to be increased, others wanted a minimum headage charge and some simply closed. It was impossible to get unanimity amongst the auctioneers and most struggled on until the next war, when many would be appointed as Government Collection Centres providing a very welcome weekly income for the company.

Sheep sale in Masham Market Square, George Atkinson of Swinton Grange, the seller in the pen
(George Chapman's Archive)

The commission charged by the marts was to remain too small as shown by an advertisement of Listers of Otley in 1934 that their selling charges were:

Each cow over £20 charged 5 shillings commission
Each cow under £20 charged 3 shillings commission
Each calf charged sixpence commission

Often total income was less than £20 per sale. To survive, auctioneers generally had to show sound judgement in their strict control of credit to buyers of fat stock with a maximum of seven days credit and the majority paying on the day of purchase.

In this period, between the two world wars, around a hundred marts closed in the United Kingdom. Much of this was rationalisation by amalgamation but this was not favoured in some areas, Yorkshire in particular, where many companies had farmer-controlled boards of directors, where at least one or two members either had that bitter dislike of their rivals, or an exaggerated pride in their own mart, which made amicable unions very difficult to achieve. Such a situation arose in Otley and is fully explained in Mervyn Lister's book, "On This Day at the Auction Mart".

Many marts were re-designed at this time as road transport increasingly took over from rail. Loading banks for one or two lorries had to be provided with manoeuvring space for the lorry to reverse in off the public road with holding pens to accommodate a load of up to some eighty sheep. A double decker Leyland or Albion lorry to hold forty lambs on each deck was something to behold.

After the major collapse of world trade in 1931, the Depression hit farmers very hard as foreign food arrived at ridiculously low prices and many English farms went out of production. The government passed a Livestock Industry Act in 1934, which gave powers to the Industry Commissioners to close auction marts and control any new ones. Fortunately, they used their powers sparingly but they did insist on closing a small number of fat cattle certifying centres; however these marts were permitted to continue to sell other stock if they wished and a few continued store stock sales, whilst others closed.

This certification was for the payment of a newly introduced subsidy on fat cattle of five shillings (25p) per cwt (hundredweight) and the equivalent of 37½p per cwt on very good quality cattle. This saved many cattle feeders from bankruptcy, with best prime cattle making £2 per cwt until World War II when more generous prices pertained

Dispersal sale of George Atkinson, Swinton Grange, Joseph Hopper presiding
(George Chapman's archive)

Joseph Hopper, auctioneer
(George Chapman's archive)

from the Ministry of Food, under the Minister, Lord Woolton, who introduced rationing of meat in March 1940. The city marts with their numerous auction companies lost out badly as they only had small surrounding catchment areas but this was mitigated by the country marts paying one third of their fees into a pool, which was distributed on the basis of a mart`s 1938 turnover of fat stock.

It was very rare to hear of a certifying officer auctioneer who did not maintain the standard of honesty expected of our profession. However, a young man in a Northern auction mart presented ten of his own cattle, which quite legally he sold privately at a decent price. His crime was telling his staff not to punch their ears so that the purchaser of the cattle could present them at another centre when the cattle would receive a second payment. The auctioneer was convicted of false accounting and fined £50, but his fellow auctioneers thought a short prison sentence would have been better.

Other government action to help agriculture at this time saw the establishment in 1934 of the Milk Marketing Board and Pig and Potato Boards. There was no help for the sheep industry.

As outlined previously, the inter-war years also saw, in 1937, the commencement of eradication of Tuberculosis from both dairy and beef herds and after the second war progress was rapidly achieved, spreading Tuberculin Attested cattle from the South West of Scotland into Yorkshire, Cheshire and Lancashire.

After World War II, several Welsh auctioneers took up jobs in Yorkshire where one showed off his perfect command of English, but when he got into full flow, he was so fast and with such a strong accent that customers would say, "he is in the Welsh again" and feared to bid as he might have been at a price quite different to their understanding. This problem was, and remains, common to a small number of young auctioneers, who tend to gabble incoherently whilst thinking themselves to be no end clever fellows!

An ability to spot a potential auctioneer enabled many auction mart owners or managers to start young lads in auctioneering at an early age. One Lancashire auctioneer told me you only have to hear a lad once to know if he has it. If he hasn't, don't waste his time, get him out of the box. I agreed with him that poor auctioneers did not improve with age.

Until 1900 most auctioneers were freelance, working for different marts each day of the week for pay of ten to twenty shillings per sale with top performers receiving a guinea, plus a free lunch and travel expenses. By 1915 rates of pay were going up and £2 was general at the end of the war and £3 to £4 was normal in the Second War rising to £5 to £8 per sale by de-control in 1954. Then wages increased four-fold in the next twenty years as prices of stock and profits increased. Pre-war profits had been small with a reasonably successful mart, such as Leyburn, reporting a profit of only £413 in 1936. In 1954 the Auctioneers Association in Yorkshire raised commission rates to 4 old pence per pound sterling for fat cattle and sheep and 6 old pence per pound for pigs (2½%) . Most firms complied, but as was usual, a few boards of directors thought it was too much.

Now most auctioneers work only for one company, although Yorkshire still has a few freelance men who now have fewer marts to go to. Salaries have risen to almost the same level as other professionals. Some command a wage of £400 per day.

Auction marts had always had competition and, as early as 1887, the Livestock Journal reported that the Lancashire and Cheshire Farmers` Co-operative Association had, in the first six months of the year, slaughtered 273 fat beasts, 903 sheep, 85 lambs, 249 pigs and 67 calves. Most of these were from farmers who were members of the Society and who thus by-passed the marts.

Co-operative organisations, then as now, were seen to be a benefit to the agricultural industry and by 1910 there were 962 such societies of which a third were dealing in eggs and a third in milk and dairy produce.

A few auctioneering firms considered joining the Co-op system but by that year of 1910 only three marts in England had converted and one in Scotland The latter, Central and Northern Marts in Aberdeenshire became one of the most successful marts, later joining with five other firms to form Aberdeen & Northern Marketing Society, today`s ANM Group. Generally auction companies did not like the idea of becoming a Co-op like the corner shop and so this form of business model did not catch on for marts. Unusually, however, the Co-operative Wholesale Society owned Gisburn Mart for a few years until 1922 when they were probably relieved to sell it to Gisburn Auction Marts Ltd managed by Messrs Turner.

After reading about Yorkshire Marts between the wars, my abiding impression is of huge numbers of prime fat cattle sold in so many un-roofed marts every day of the week, but predominately on Mondays, and every week of the year. Entries of 500 cattle were common and, even allowing for those sold twice in a week, first at a country town then at a city mart, it demonstrates what a fantastic feeding area Yorkshire must have been. Driffield and Malton each sometimes had a thousand fat cattle on one day, more than are now sold at every mart in Yorkshire, Lancashire and Cheshire combined. I am sure the feeders will come to regret their loss of their auction outlet. They have over-supported the direct sales alternatives and have thus lost the competitive edge provided by open market pricing.

CHAPTER XIII
SOME AUCTIONEERING FAMILIES

As the first century of auction marts progressed, a few families had, by their innovative minds and good reputations, become firmly established as hereditary auctioneers. Most left little record of their exploits or ever owned a mart and, worse still, in the last twenty years many young men from those families have chosen to take up different occupations.

However, a few have kindly shared their recollections with me and this chapter tells, at widely varying length, of the marts of Mervyn Lister, James Stephenson, the Cundalls and the Kirbys from Yorkshire and Richard Turner from Lancashire. Mervyn Lister has written a wonderful book on his family marts and the fathers of James Stephenson and Richard Turner served with me on the Livestock Auctioneers Market Committee for England and Wales in the 1970s, where they played an influential and forceful part. I also mention the Ivesons who were friends of my father and who, even if more famed at Hexham, had strong roots in Yorkshire.

The Turners

Richard Turner, son of a Craven farmer, saw a fair lass and married her in 1828. Her father was a general auctioneer in the Bentham area who needed help at his sales and so Richard became an assistant auctioneer and his son, also Richard, became a most successful livestock auctioneer. Then the third Richard set up Bentham Auction Mart Ltd and built a good new mart there in 1903, a century after the occasional sales had started and ever since it has been a flourishing centre with a reputation as a leading sheep mart.

In 1952 the next Richard, whom I knew well as a highly respected auctioneer whose ability and integrity was well known over a wide area, took control of Gisburn mart and expanded the business in a very short time to be one of the leading dairy cattle marts in the country, where he sold every week until he was over eighty. He also built up a large estate agency business and died in 2008. He would be pleased that the next Richard continues to expand the business assisted by his son, Jonathan and a good team of auctioneers.

Fat sheep sales continue to be of immense importance to the

farming community, but they are constantly under pressure and in February 2016 Gisburn mart closed their Monday sales of prime sheep and added it to their Thursday sales of cattle, with expectations of substantial numbers. An annual show is held here on the first Thursday in March. Prime sheep continue to be sold on Wednesday evenings in Bentham, which is one of the larger cull ewe marts in England.

"R. Turners Auction", circa 1951

For a hundred years the Turners held several popular sales each year in various country towns and villages including Ribblehead, Slaidburn, Ingleton, Kirkby Lonsdale and Settle. These have now gone but their farm displenishing sales are still held over a wide area of the North of England.

The Stephensons

Amongst the early auctioneers around York were the Stephenson family, who had already been involved as appraisers, valuing for the Fairfax family at the end of the Civil War in 1645, when they moved from Steeton Hall to Newton Kyme, two villages either side of Tadcaster.

By 1871 Jacob Stephenson was living at Manor Farm, Wighill and opted to set up his board in opposition to Matthew Thomlinson at Tadcaster Market some five miles away. But the promised land of York beckoned and, in 1900, Jacob, with his son, Richard, moved to the Corporation owned market there to pay their tolls for pennage along with eight other auctioneers.

It was a blow to the Stephenson firm when both Jacob and Richard died in 1927 leaving the next generation, Fred, together with his twenty year old son, Reg to fight for business. Reg was not only a gentleman, but he also had steely determination, vowing at an early age that he would make £1,000 per annum by the time he was 30 years old, a goal which he achieved.

An office was set up in the centre of York above his father-in-law's restaurant at 31 The Pavement; and the firm continued to sell at both York and Tadcaster with auctioneers and staff taking the train to the last stop at Bolton Percy before walking the last four miles into Tad!

1939 saw the outbreak of war, the closure of livestock auctions and introduction of control whereby all the animals were brought to the market premises, graded by a panel of one farmer and one butcher, before being allocated to wherever the meat was needed.

The meat shortage continued long after the war finished in 1945 and it was not until 1954 that the Government finally released its control allowing auctions to start up. By this time, there were only four firms left at York – R.M. English, Joseph Cundall & Sons, Don Johnston and Stephenson & Son.

The Stephenson story is inextricably linked with that of York Market and readers are referred to the section on York and district in Chapter Five of this book. Suffice to say that, without the drive and enthusiasm of Reg Stephenson, there would not be a mart in the city today.

The new York Livestock Centre, a joint venture with R.M. English, was opened in 1971. Reg Stephenson shared and then handed over the selling duties to his sons, Nigel and James, who, together with their exceptional head auctioneer, Cliff Lockwood, and John Burley of English's, rapidly built up trade over the next decade.

By 1983 York was second only in scale to Banbury and for five consecutive weeks in March had over 1,000 head of fat cattle, but perhaps even more unusual was the growth of the pig market, which Cliff and Nigel pushed to sell through the live ring 4,000 pigs in one day. Pigs would start arriving by 4pm on the Sunday afternoon and the sale commenced at 8am on Monday morning, with every pen being filled three times by incoming replacements. From those heady days, numbers have been decimated by successive "notifiable diseases" such as swine fever, blue tongue, BSE and Foot & Mouth. With each outbreak the auctions were temporarily closed and this forced farmers to source direct purchasers for their stock. The 2016 figures for York show around 200-250 prime cattle per week with some 500 pigs and 1,000 sheep. To fill the void, diversification and salvation was found in resurrecting the machinery sales, which are now the largest in the North of England, with 5,000 lots in each of the nine monthly sales.

Sadly, Nigel Stephenson died in 2004, but his brother James still heads the firm. Stephenson & Son remains a family partnership, now in its sixth generation, boosted by the additional talents of Richard Tasker, who masterminded the machinery sales, Bill Smith who came from Carter Jonas following Nigel's death and, more recently, Davina Fillingham, a Nuffield scholar and RICS student of the year. York Market has and will continue to be the shop window for the family firm, which enriches the entire community.

The Listers

Joe Lister and his son, Fred, started a second auction sale of livestock in Otley in 1890, where Dacre & Son had been selling at the Wharfedale Mart since 1870, and later extended their business to include seasonal sheep sales at Ilkley and Fewstone. Fred was a popular auctioneer with large sales in Otley, but he also sold at Skipton, Hellifield and Bingley marts. This selling for other mart companies was the custom in Yorkshire marts whereas in Scotland, auctioneers only sold in their own company's marts.

The severe depression after the Great War forced the firms of Lister and Wharfedale to join together, in 1922, at the East Chevin Road Mart under the name of Wharfedale Farmers Auction Mart Ltd, with Fred Lister as chief auctioneer.

Fred continued to sell at other marts until inevitably the problem of a conflict of interest arose. When John Smith moved his consignments of quality cattle from Otley Mart to Bingley, somebody must be responsible so, as I always found in fifty years of auctioneering, the auctioneer was blamed! The dispute came to a head in 1930 when Fred accepted the post of chief auctioneer to the Bingley Livestock Auction Mart Co Ltd. He had previously agreed not to sell at any

YORK ATTESTED MARKET

Sales of Pedigree, Supplementary and Commercial Dairy Cattle from Attested and Supervised Herds held on alternate Thursdays in every month.

Stephenson & Son, F.A.I.

20 CASTLEGATE, YORK

Tel. 2513

Auctioneers and Valuers of Real Estate

Tenant-right and Live and Dead Farm Stock ; Management of Estates and Surveys of all descriptions. Monthly Sales of Agricultural Horses and Farm Implements, Tractors, etc.

mart within eight miles of Otley. Fred said that it was nine miles to Bingley and the Wharfedale directors contended that it was only eight.

Fred Lister, Willie Lister, F Mervyn Lister

The court case and the results are reported in the following description provided by his grandson, Mervyn Lister:

An extract from the Yorkshire Evening Post of November 12th 1930
"A SUPREMELY GOOD SALESMAN"
Two Auction Marts Move Towards Settlement of Otley Case

"An agreement's bearing on the professional activities of an auctioneer who was described as a supremely good salesman formed the subject of a case at Otley County Court today, in which a radius provision was involved, and the question of the distance between Otley and Bingley was raised.

The Wharfedale Farmers Auction Mart, Co. Ltd., Otley, claimed an injunction against their chief auctioneer, Mr Fred M. Lister of Askwith, to restrain him from acting in a similar capacity at Bingley, and £10 damages. Mr Raymond Hinchliffe (barrister) represented the plaintiff company, and Mr H.C. Davies (Leeds) was for Mr Lister.

Mr Hinchliffe said the Auction Mart Company was formed in 1918. At that time there were two auction marts in the town, one conducted by the late Mr J.W. Dacre, and the other by Mr Lister. The company bought Mr Dacre's business and arranged to carry on their business in his premises. The defendant's business was not then in a very prosperous state, and it was ultimately agreed that Wharfedale Mart would buy it for £4,500, included in which was £1,500 for goodwill. Mr Lister was engaged as chief auctioneer at a salary of £250 per year, and included in his agreement was a clause to the effect that he must not be interested in any other similar business within a radius of eight miles.

No objection was raised to his acting as a valuer, and he now had one of the principal valuing businesses in Otley. He also acted as salesman at Hellifield and Skipton Auction Marts, and to this there could be no objection, as they were outside the radius.

In June this year it came to the knowledge of the plaintiff company that the defendant was conducting sales at Bingley auction mart, which was in serious competition with the Otley mart. It was also learned that he had entered into an agreement with the Bingley mart to serve as chief auctioneer."

Fred could not sell within eight miles of Otley and Bingley was nine miles by road, but only eight as the crow flies. The case angered many of Fred's friends, who set his son up in a new mart at Otley in the old Market Square, to be known as Otley Cattle Market Auction Ltd and to operate in competition to the Wharfedale company. Fred stopped selling at Bingley.

The group of farmers who wished to promote this alternative auction company were keen to see some better equipped and more comfortable premises provided for the purpose and, during 1931, they had decided to form a new company named Otley Cattle Market Auction Limited and it was incorporated on 30th November 1932 with William Walker Lister (Fred's second son) as the company's secretary. It then needed some premises from which to operate. The new company (O.C.M.A.) had its registered office at Barclays Bank Chambers, Manor Square, Otley. Its authorised share capital was £1,000, and in total some £4,600 was raised by numerous farmers and butchers in the area to fund its launch.

Although at that time only 23 years old, William (or Willie, as he was more commonly known) had, by then, already been working for almost ten years in the family business of auctioneers and valuers. He was newly married and was studying for the professional examinations of the Chartered Auctioneers and Estate Agents Institute.

In 1931 the Otley Urban District Council agreed to let O.C.M.A. operate in their "off street paved area" in Cattle Market Street, for the purpose of conducting auction sales of livestock in a temporary building, which the new company would erect on that site. The first sale took place there on 20th January 1933, and gradually, but steadily, the throughput of that business began to increase. Soon the aspirations of O.C.M.A. encompassed the wish to have the comfort of owning their own premises, which they could develop to a higher standard. Within another year or so they purchased a field next to Otley Bridge on the north side of the river. It was known locally as "The Drill Field", presumably because it had been a place where the young men entering the armed forces in earlier years had been trained. It was two acres in area and was bought from Fred Lister for £255.

Wharfedale Mart, Champion Limousin Cross Steer from Brown Bank Farms, April 2016

During the following year, Oswald Holmes & Son, a local firm of architects, drew up plans for new buildings on the site and construction work commenced. Unfortunately, there was a massive flood during February 1935 at which time the new buildings were "half built". Some desperately bad economic circumstances of the day resulted in the firm going into liquidation. Willie Lister and his stalwart farming colleagues then worked on the site themselves for several months in order to complete the project. The premises, Bridge End Mart, were sufficiently advanced by midsummer 1935 for sales to commence on 20th June that year. William Lister was the auctioneer that day and was the auctioneer and company secretary of O.C.M.A. for many years. From that first sale, under Willie and Tom Lister, the Otley Cattle Market Auction Co Ltd., slowly, but steadily expanded business until 1940 when all fatstock marts became Collecting Centres for the Ministry of Food and continued in that role until 1954.

Wharfedale Mart, Calf Sale

However, during this period there was a revolution in the dairy cattle business at Bridge End Mart, which was one of the first Yorkshire marts to be approved for selling tuberculin tested cattle and grew swiftly to be one of the most important dairy cattle marts in England with stalls for 500 attested cattle and progressing from selling Shorthorns to Ayrshires and Jerseys and finally to Friesians.

In 1952 Mervyn Lister joined the team bringing some new ideas, and continuing to develop the firm for over sixty years. The next year, 1953, Bingley Auction Mart was bought for £1,000 and Listers set to work to rebuild it ready for decontrol in 1954, when it became a most successful centre until its closure in 1996. In 1978, Mervyn Lister was joined by his son, Martin who later sold at Thirsk and continues to be involved with Clitheroe Mart. His brother Graham joined the firm in 1990.

Bridge End prime stock trade increased and, with the closure of Pannal and Bingley, compensated for the reduction in throughput of dairy cattle. Some important breed sales were held in conjunction with Thimbleby and Shorland of Reading for Ayrshire cattle and with Hobsons of London for the British Friesian Society.

Dairy cattle sales became much smaller as recorded in Chapter Eight and, in 2000, the last Lister's mart closed. They continue as surveyors, land and estate agents.

Wharfedale Auction Mart

The Ivesons

Arthur Iveson was destined for the Church but instead became a successful cattle dealer based at Hawes in the mid nineteenth century. His sons, William (1852 – 1922) and Thomas Taylor (1854 – 1934) were among the most respected auctioneers in the North of England taking over William Cook's mart in Hexham in 1879. They established marts in Hellifield, where Thomas Taylor was chairman for many years, and in Hawes in Yorkshire where they traded as Iveson and Pratt, later to become the Hawes Auction Mart Ltd. They also had a very successful mart at Bellingham in Northumberland, where 20,000 lambs would change hands at their two day sale. Later they would take over their competitors in Hexham, the Tynedale Auction Company, whose strengths had been Irish cattle and dairy cattle, whilst the Ivesons conducted the largest store cattle sales in England every Friday. On one memorable Friday over 4,000 cattle were sold. In 1991 they bought the Scots Gap Mart and Hexham and Northern Marts continue to hold sales there.

William's two eldest sons were killed in the Great War. As well as the mart businesses, the family also carried on the private firm of W. & T.T. Iveson until finally fully merging it with Hexham & Northern Marts. William and Thomas were succeeded as managers by William's son, Michael and grandson, Eric, and Thomas Taylor's sons, Arthur and William. Each had a son, Arthur and Billy, who were managing directors from 1967 until 1996. During this time they planned and built a very fine new mart at Hexham. With over five hundred pens and three sale rings, this was opened by Lord Plumb in 1995. At the opening, there were three Waltons and two Ivesons on the board of Directors as well as the popular Maurice Reid, who took the opportunity, as he went round farms, to make a great collection of agricultural memorabilia at a time when much of interest was being thrown out. Reid was also a very shrewd valuer as I found out when doing sheep stock valuations. Arthur Taylor Iveson's grandson, Robert Addison is now managing director operating a very successful business. Hellifield has closed but Hawes goes independently marching on.

The Cundalls

Joseph Cundall was a farmer, cattle dealer and valuer living at Sherburn midway between Malton and Scarborough. He started in business in 1860 and opened his first weekly auction sales in 1880, moving his sales to Malton in 1885. It is said that he, and later his son, would walk the thirteen miles to chapel in Scarborough each Sunday morning and home again for a late lunch!

His business at Malton grew to be one of the largest fat cattle marts in Britain with his grandson, John, selling 1,640 prime cattle in one day sometime after the Second World War. Cundall`s sold alternately with Boulton and Cooper and, when "first turn" would often sell over one thousand cattle as late as 1980. John Cundall was one of the outstanding auctioneers of his day, respected by all. He died in 2010 aged 83. His brother Harry, also highly thought of, managed the firm`s land agency business for many years. As with many Yorkshiremen, horses were a great enthusiasm of the Cundalls, with John`s son Chris, now a veterinary surgeon, proving to be a successful National Hunt jockey. James, another of his sons, is a distinguished and successful theatrical producer. Sadly, or perhaps wisely, none of his sons followed him into the business. Tom Watson at Cundalls continues to sell many Yorkshire farms and, in 2017, handled the sales of almost 5,000 acres.

The Kirbys

Seth Kirby was appointed senior auctioneer when the Thirsk Farmers Auction Mart Ltd was founded in 1907. He had started as a livestock auctioneer at Helmsley in 1889, where he later took on Gerald Perriam as a partner, also doing a day at various other marts and establishing Autumn sales of sheep at Rosedale Abbey, Bilsdale, and Hawnby on the North Yorkshire Moors, and establishing an office at Ripon where his son, Ernest, succeeded him as chief auctioneer in 1921, on his return from the army. Seth had four sons all of whom joined the auctioneering and valuing business, selling regularly at Helmsley, Ripon, Northallerton and Thirsk, as well as helping out at other centres.

Ernest Kirby was senior auctioneer at Thirsk and Ripon for 33 years and died in 1964, a much-respected member of the profession having been president of the Yorkshire and Durham Agricultural Valuers Association. 1965 saw the firm of Seth Kirby & Sons take over Thirsk Mart from the Thirsk Farmers Auction Mart Ltd., only for Thirsk Farmers Mart Ltd to take back the mart in 1972 and continue to run it but with the Kirbys very much involved. In 1988 Boroughbridge Farmers Auction Company was taken over, but it was a difficult time and this was soon closed down to concentrate on Thirsk, with Ripon also closing, as so many marts did after the outbreak of Foot & Mouth Disease in 2001. Ernest`s son, Tom, was a director and secretary of the Thirsk Auction Mart Company until he retired in 2014. The family had spanned 125 years in the marts.

CHAPTER XIV
THE MARTS OF LANCASHIRE

The appendices to this book contain a summary of all the marts operating, at one time or another, in Lancashire and details of such dates and auctioneers as have been traced. Inevitably information is sketchy. The profusion of local histories which have been produced over the years generally give no mention of the local cattle market and no hint of the people who worked and traded there. Likewise early photographers seem to have gone out of their way to avoid getting their feet dirty or their camera apparatus damaged among the cattle pens. Here, in alphabetical order, is an attempt to record some brief facts and information about a selection, but not all, of the marts which are listed later.

A new cattle market was erected in Harrison Street, **Blackburn**, in 1874, with public slaughterhouses adjoining. An Eli Smith was an auctioneer here in 1876, followed by Whittaker and Armitstead in 1880 and Blackburn Farmers Mart in around 1890. Blackburn like many marts in industrial towns was designed to supply the local butchers and milk purveyors. It was reputed to have been well designed and built. This was borne out fifty years later when Henry Manley of Crewe would use it for dairy cattle sales including special sales of up to sixty Ayrshire milk cows and heifers from Robert Morton, an Ayrshire cattle dealer who bought sheep and cattle at our Thornhill mart; he was a dapper, educated, man who was a pleasure to do business with. This small mart survived for almost a hundred years, but after de-control in 1954 it struggled on and then closed, almost unnoticed, in 1990.

In **Bolton** a council-owned cattle market was opened in Orlando Street sometime after 1900 but no regular weekly auctioneer is recorded here. Periodic sales were, however, held at Horwich, approximately three miles away with John Rodwell in the rostrum from about 1910 to 1930.

Bentham had a mart from the early date of 1840, the controlling influence here being the Turner family (see the previous chapter). This remains a thriving centre with large regular sheep sales and it is a real hub for the agricultural community.

At Claughton on Brock, the **Brockholes Arms** stands beside the traditional drove road where the Scottish cattle would cross the Wyre and continue their journey south. Auction sales here go back to the 1880s with J. Kelsall the auctioneer in 1885. In 1927, The Brockholes Arms Auction Mart Ltd. was formed by a consortium of local farmers who leased the mart site from Major Fitzherbert Brockholes, later, in 1948, the firm was able to purchase the site outright from Major Brockholes for £450. In the late 1920s, the Garstang Auction Company had moved their modest business to join with Brockholes. **Garstang** had been described by John Ogilby in 1690 as "a good thoroughfare with a market for corn, cattle etc.on Thursdays". The mart there was, however relatively short-lived with a lifespan from 1880 to about 1927. In 1985, T. Armitstead & Sons (now Armitsead Barnett) became auctioneers and secretaries to Brockholes Mart and, in 2001, the Brockholes business was greatly enhanced by the closure of Preston Mart with Brockholes changing their main market day to Preston`s Tuesday. Brockholes Arms remains a well-run farmer`s mart with a very supportive board of directors.

In 1866 the Borough of **Burnley**[33] were granted a licence by the Privy Council to hold a cattle market in Parker Lane and, because of the risk of spread of the Cattle Plague it was stipulated that all stock must be slaughtered within four days of sale and must be slaughtered within the Borough boundaries. The market was to be held each Thursday for cattle, sheep, goats and swine but for slaughter stock only. In 1867, there were announcements in the Burnley Gazette that "the whole of the cattle brought to Burnley were sold at fair prices" and that "dealers at a distance should understand that we can now furnish a good market for their living wares". Some twenty years later, in 1886, a new mart was erected, the auctioneers being Ackroyd and Green. In November of that year, the local newspaper was reporting that:-

The new cattle auction mart of Messrs Ackroyd & Green is now completed. The structure is built entirely of wood with a roof of corrugated iron. It is divided into two portions, one containing the cattle pens and the other the place of sale. The latter is fitted up with a rising gallery for buyers, an open space being left in the centre for the cattle on sale.

Surely a good and simple description which would have fitted so many marts of the time. Ackroyd and Green held fortnightly sales on a Monday. Perhaps by 1890, the mart was struggling as Mr Ackroyd

33 Edward Walton, of Walton and Walton, formerly chattels auctioneers in Burnley, has assisted greatly here.

wrote to the town clerk requesting to be "released from his tenancy of the cattle auction mart". The council replied that this request could not be entertained and so Ackroyd and Green doubtless soldiered on. By 1907, a Herbert Lee of Colne Road, Burnley was advertising sales in the Burnley Cattle Mart, including small sales of dairy cattle. The site of the mart is now occupied by the Police Station and Law Courts with the Cattle Market Inn across the street in Parker`s Lane. The mart closed in 1920. There were also small marts in the surrounding villages including Barrowford Cattle Mart, on the south side of the Gisburn Road, Cliviger and Blackshaw Head. At each of the two latter locations, Walter Taylor of Burnley held monthly cattle sales.

Gisburn the last Fair, 1911

Clitheroe remains a busy centre with an innovative sales programme under Joe Worthington and Jeremy Greenhalgh with Robert Parker as managing director. The Clitheroe Auction Mart Company Limited was originally established in 1897, specifically to take over the ongoing business of Samuel Green of Clitheroe, "Auctioneer and Cattle, Hay and Straw Dealer". Initially there were 5,000 shares of one pound each and, as usual with marts, the promoters were a combination of local butchers, farmers and cattle dealers, in this case with the auctioneer

Samuel Green also being involved and a local accountant as the Company Secretary. It proved to be a successful enterprise and they also operated the mart at **Chipping** from 1901 to 1955. Ninety years after their foundation, in 1988, a well-planned new mart was opened in Clitheroe by Baroness Trumpington, Parliamentary Secretary at the Ministry of Agriculture. The Clitheroe company name remains unchanged today. The firm of Silverwood and Balshaw were also based in Clitheroe and, in 1995, they were acquired by The Clitheroe Auction Mart Company who absorbed Joe Balshaw`s cattle business into their own but retained the Silverwood`s name and business as their fine art and furniture division.

Gisburn, too, still goes strong. On 21st November 1260, The Abbot and Monks of Salle were granted a Royal Charter which entitled them to have a Market every week on Monday at their Manor of Giseburn in the County of York and a Fair of three days duration every year. The last of these street fairs was held in Gisburn on the 16th October, 1911 as it was being superseded by the first auction mart built by Lord Ribblesdale, part of the original building still being used today. On the 18th May, 1911, the mart was leased by Lord Ribblesdale to Rowland I. Robinson, Land Agent, Henry Jackson, Auctioneer, and Henry Charles Starkie, Farmer, at a yearly rental, plus, as royalties, one penny for every head of cattle, fourpence for every score of sheep, one penny for every pig, and sixpence for every horse, which passed through the ring. Later Messrs. Jackson and Starkie purchased the premises and they, in turn, sold it to The Co-operative Wholesale Society Ltd., who ran it with Edgar Jackson as auctioneer until 1922 when they sold it to a group of gentlemen comprising Messrs C. Bullock, R. Campbell, J.W. Gill, E. Johnstone, H.C. Starkie and A. Tattersall. It then became Gisburn Auction Mart Ltd., and in 1948 Skipton Auction Mart Ltd., purchased a half share and traded as Gisburn Auction Marts Ltd., under which title it still continues. Business was rather neglected and in 1952 the company sold out and a new board of Directors, all men of good farming stock, took over. They were Messrs A. Pickles (Chairman), J. Aldersley, G.W. Bargh, H. Maude, E. Drake, J.T. Pratt, and H.F. Taylor. They appointed Messrs. Richard Turner & Son of Bentham as Auctioneers and Secretaries and there was an immediate upturn in throughput. Turnover in 1952 was £130,662 and this rose to £2,000,000 by 1959. The throughput in that year was:

Newly calved cattle	10,894
In-calf Cattle	4,617

Store Cattle	1,948
Attested Bulls	152
Fat Cattle	7,159
Graded Sheep	42,149
Other Sheep	11,409
Graded Pigs	5,312
Store Pigs	15,473
Sows	1,162
Calves	10,272

In 1958 a large aircraft hangar was purchased and converted into an excellent building for the fatstock ring, calf ring, and shippons capable of holding 300 cattle and 400 to 500 calves. Around these rings and the dairy ring, through which for many years passed on a Thursday an average of 300 newly calved cows and heifers, can be heard dialects from all parts of Yorkshire, Lancashire, Westmorland, Cheshire and from much further afield. This positively run business, with Richard Turner as chairman and Stephen Dennis as senior auctioneer continues to be a leading player among the marts of the UK.

Haslingden Mart, the Board of Directors, 1950

Haslingden had a mart since 1890 with the Haslingden Cattle Market Co. being established in 1920 with eight shareholder directors. The mart was at Bentgate, near to the cricket ground. From the 1950s, auctioneers, Norman Duxbury, his son Ted, and Albert Mawson had a large and very loyal group of supporters here with good numbers of fat, dairy and store stock coming every week. Joe Worthington and Mark Schofield kept a good business going until B.S.E. spoiled the party. Billy Pilling had shown fat cattle every week and there was always a wagon load from Wishaw Mart in Scotland. Charles Fieldhouse bought cattle every week as did a number of other loyal wholesale and retail butchers. Ultimately Andrew Birtwhistle, a successful Manchester meat wholesaler became Chairman and bought the business. At this time the value of the eight shares was £30,000 each. The capital value of the site tempted him to take a substantial profit and the mart closed in 1986.

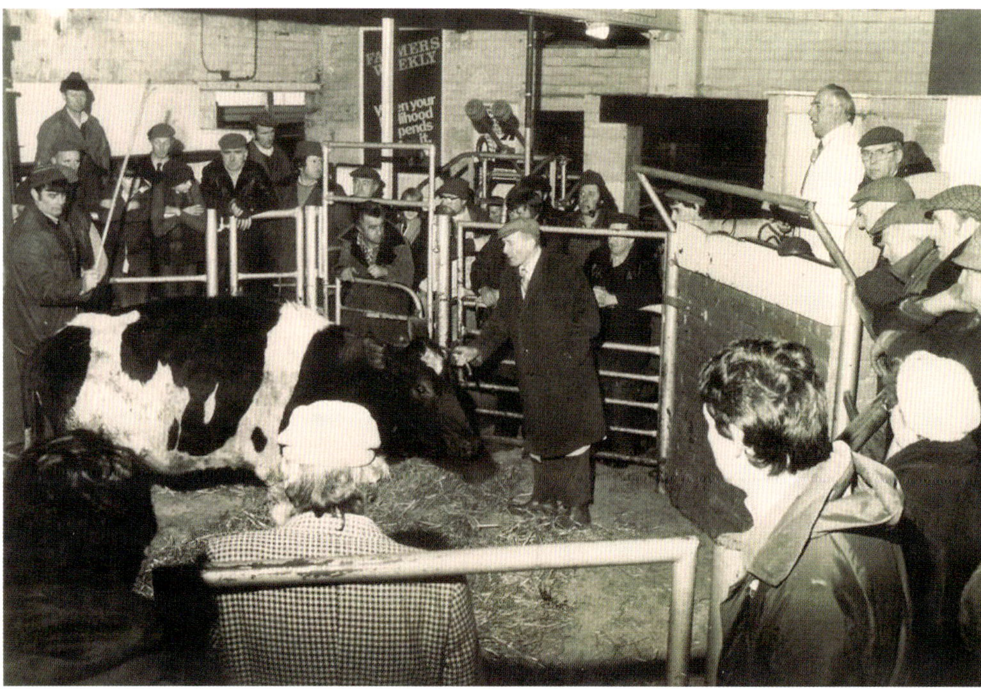

Haslingden Auction Mart

In 1193 **Lancaster** received its first charter enabling it to hold weekly markets and an annual fair. The town received mixed reviews from early 18th century visitors with Daniel Defoe telling us that it "had little or no trade and few people"; others, however disagreed and, in any case, it developed hugely in the late 18th and, again, in the late

Lancaster Auction Mart 2017, Ian Atkinson auctioneer

19th century with fairs for "horses, cattle, cheese etc." being held on 1st May, 5th July and 10th October; the October date was the principal cheese fair in the county. The Prince William Henry public house, later the Corporation Arms, in Penny Street had an adjoining field where, by 1880, the "Lancaster Auction Mart" was advertised with fatstock sales, on a Tuesday, by W. Haythornthwaite who also sold at Ulverston and elsewhere. He had a hide and skin sale in Lancaster on a Friday and, some Saturdays, sales of store cattle and quarterly horse sales at this "Prince William Henry Auction Mart", to which he charged half a crown for admission. Mr G. Corbishley of Thurnham, was also very active as an auctioneer by 1880 and sold fatstock at the "Wheat Sheaf Auction" each Wednesday as well as conducting many sales on the sellers` premises. In 1885, a mart was erected in Thurnham Street and the Lancaster Farmers Auction Mart company was established in September of that year. James Downham, of Alexandra Road, Morecambe, also sold in the mart, both on his own account and, later (by 1935), jointly with the Farmers Auction Company. Thomas Armitstead, too, sold in Lancaster and elsewhere and acquired Corbishley`s business in 1890. Horse sales were very much his speciality; and his family name continues to be involved with auctioneering in Lancashire as Armitstead Barnett operate Brockholes Mart today. At the time of the Lancaster mart`s erection Thurnham Street was a cul-de-sac but in 1901 it was opened up and continued through the Prince William Henry Field where sales had previously been held. From before 1900 regular sales of consignments of sheep from Scotland were advertised and the author recalls that, until 1940, Peter Hyslop and Sons of Annan would send large numbers of Scottish store sheep, some their own and others for which they were agents, fortnightly to Lancaster Mart. The Armitsteads, as Thomas Armitstead and Sons continued to sell in Lancaster Mart; firstly Thomas`s sons, Reg & Herbert (although Herbert died untimely) succeeded in turn by Thomas John Armiststead who was in partnership with Jacky Johnston and followed latterly by Geoff Armitstead who also ran Pilling Mart. For many years Lancaster was a major sales centre for British Friesian cattle, under the auspices of Hobson`s of London. The Thurnham Street Mart survived until 1965 when the present premises were erected on Wyresdale Road. In 2003, there were proposals for the formation of a partnership involving Harrison and Hetherington and the Kendal Auction Co. but this came to nothing and subsequently, in 2005, the Lancaster and Kendal businesses merged with a new mart being built in 2013 at Crooklands, Junction 36, of the M6 and trading as North West Auctions. They still successfully operate Lancaster Mart today with substantial sales of store cattle and continuing sales of dairy cattle and primestock.

Preston was a progressive town where large horse and cattle fairs had been held since Norman times. It was one of the earliest providers of a fenced market area and had a farmer in control at this council owned market. In 1867 E.G. Hothersall started regular auction sales here and his sons, John and Wilf, carried on one of the largest family marts in England with John Kelsall, a popular auctioneer. Large dairy cattle sales were held on Monday and Friday with fatstock on Tuesdays and Wednesdays. The mart was adjacent to the railway station with the Hothersall name very prominent for all travellers to see. In 1935 Hothersalls were advertising October sales of 120 smart in-pig gilts and 150 horses but their most profitable trade was a weekly cast cow sale, which was one of the largest in the country and carried the mart forward until B.S.E. controls in 1996 terminated this aspect of the business. In 1983 Preston had joined with Kendal

Mart and they ultimately closed in 2001. For many years weekly sales were held at Kirkham, Longridge and Poulton-le-Fylde, but after 1954 motor transport took the business into Preston.

Edward Ashworth opened a small auction mart in the 1920s/1930s to serve the **Rawtenstall and Bacup** area. He was assisted by his two sons who also had an abattoir for the convenience of some local butchers. During the war it was not a government grading centre but they continued to sell dairy cattle bought in Scotland by themselves for resale. From 1954 fat stock sales commenced and continued for about ten years, while sales of dairy cattle went on until 1979. The abattoir was sold in 1986 and their weekly importation of Irish fat cattle ended after seventy years and three generations of their family involvement. They had been driven out of business, as were many of their fellow wholesalers in the industrial belt of Lancashire and Yorkshire, by the supermarkets who only took meat from the largest abattoirs. Harry Ashworth still takes an interest in the agricultural scene.

At **Warrington**, historically, until 1974, in the county of Lancashire, auction sales were initially held in the inn yards of the Eagle and Child and The Lion hotels and, a little later, at the Norton Arms. The Eagle and Child sales in Warrington were held by John V. Mainwaring; his firm later became J Mainwaring, Stafford & Co. and was later owned by Stanley Wright. By 1895, Warrington town council realised the need for a proper market site and an acre at the old Militia Stores, near to the railway station, was procured for the purpose. It was used by several auctioneers and, by 1900, W. Bainbridge of Froghall Lane had an extensive business including horse sales. The Norton Arms Mart must have continued for many years as, in 1939/1940, Marsh and Son of Nantwich were still advertising that they were at the Norton Arms Cattle Market on Mondays of each week. Marsh`s, established back in 1875, continued to trade in the person of Donald Marsh and they, and the Mainwaring business were, in due course, acquired by Frank Marshall of Chelford. In **Wigan,** the Ship Hotel was an early sale venue and various premises accommodated auction sales until, in 1937, a fine new mart was built in Prescott Street, Wigan, near to the railway, with pens for 160 cattle, 200 calves and 1,000 sheep and pigs. John Trickett had acquired the business in 1901 and, by offering payment on the day of sale, built up a successful business including store stock and estate agency. The Trickett family and their connections managed the Wigan business for some ninety years. Warrington closed in 1960 and Wigan in 1992.

CHAPTER XV
THE MARTS OF YORKSHIRE

As with the previous chapter, a fuller summary of the marts and their auctioneers may be found in the appendices to this book. Here is some information on a selection of the county and provincial marts of Yorkshire.

Situated between Bradford and Leeds, and very convenient for the butchers of these cities to attend, sales were held from about 1870 in the Queens Hotel Yard at **Apperley Bridge** by Thomas Burroughs. By 1890, W.H. Pountney was selling from the Stanhope Hotel yard and, by 1890, it is recorded that Mason and Jennings, on a Monday in September, were offering 100 fat cattle, 500 sheep and "twelve splendid shearling rams from Mr Langley`s flock". These auctioneers were succeeded by the Apperley Bridge Livestock Auction Company and the mart here ultimately closed in the early war year of 1940.

From as early as 1845, a Mr Wilby held sales in the village of Denby Dale, about eight miles from **Barnsley.** By 1860 the Barnsley council were building a market with Wilby selling there. Pigs were the subject of a weekly auction with prices at that time of about £1 per head for young pigs and baconers to £4. A large fair was held each October and, from 1890, regular sales of imported Jersey and Guernsey cattle took place. Lancaster and Sons were selling from 1890 and Wilbys continued trading through the erection of a new mart in 1933 (when C.E. Smith & Co. commenced sales also) and on through to the 1960`s when, in 1961, the weekly sale embraced around 100 cattle, 225 sheep and 135 pigs. The mart closed in 1990.

By 1880 several auctioneers attended in the town of **Bedale** and sold livestock. Then, in around 1900, some local farmers established the Bedale Auction Company which held weekly sales of fatstock until 1935 at which time Oswald Lister bought their goodwill and carried on the market until 1960. It is believed that he didn`t operate a Government collecting centre during the War Control period to 1954. George Wetherall and Roy Taylor then conducted the business until 1971 when they sold out to Norman Brown and his brother-in-law Gordon Pratt from Hawes. They modernised the mart and sold about 250 fat cattle and a 1,000 prime sheep at their weekly sales until the very early 1990s when the site was sold to the County

Council. Gordon went to manage Hawes Mart and Norman retained the office and carries on a successful land and estate agency business in Richmond and Bedale which recently celebrated its fiftieth anniversary.

A Pig Sale at Beverley

Beverley, in 1727, was one of the earliest towns to obtain an Improvement Act that included providing a new market area in 1727. From 1860 auction sales were held and by 1935 eight firms were holding weekly sales. At its peak, it was a very large pig market with over 2,000 head of fat pigs in many weeks, and sometimes as many as 5,000 store pigs. Latterly some of the Hull business was relocated to Beverley but the mart succumbed, on 1st April 2001, to a purchase offer by Tesco. Some of the business transferred to the new "Hull" mart at Dunswell, about seven miles away.

Tillotson and Son were selling in **Bingley** by 1890 and, in 1906, a William Ash of Garforth acquired the lease on payment of a sum of £500. Bingley Livestock Auction Mart Company was established in 1914 with Bernard Bartle as secretary. They purchased the freehold of the site in 1920 and in 1922 Fred Lister was appointed auctioneer. New pens and railway sidings were erected in 1930 to serve the Irish cattle trade and after the war, in 1953, the Otley Cattle Market Company, managed by William Lister bought over the mart. The old premises were rebuilt and business developed. In 1935, a typical entry would have been 150 fat cattle and 600 fat sheep but by 1965 this would have increased to about 250 cattle and 800 to 1,000 sheep, together with some 350 pigs. Bingley Mart, in Keighley Road, closed in 1996.

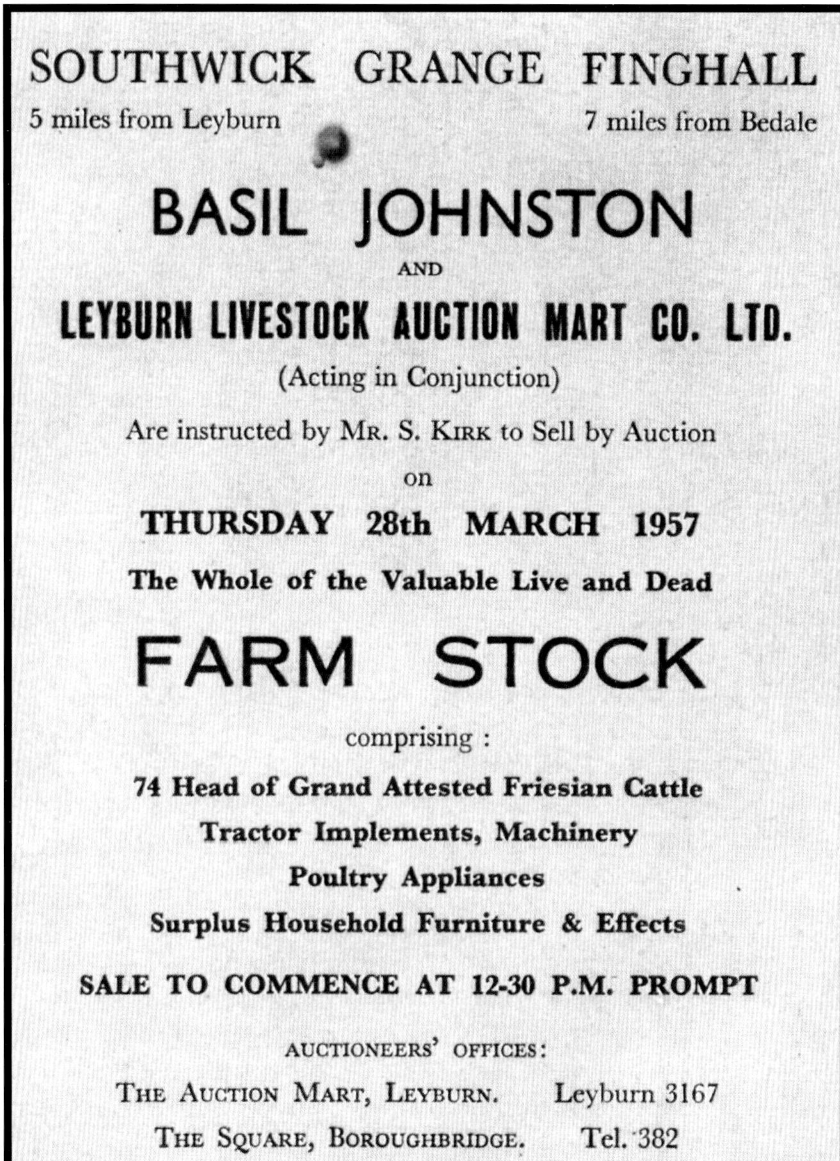

Boroughbridge, home to Barnaby Fair, one of the largest fairs in Yorkshire, stands on the A1, some twenty miles north west of York and was, for over eighty years, home to the Johnston family of auctioneers. From about 1900, they had attended several Yorkshire marts to sell their clients` stock but, by about 1920, James Johnston took the view that his two sons would not work together well and placed Don at York and Basil at Boroughbridge where the Boroughbridge Farmers Auction Company had sold on Market Hill since about 1910. Both Don and Basil were popular, entertaining and successful auctioneers. In 1935, 2,600 breeding ewes and lambs were sold at the annual September Fair. In 1960 average weekly numbers were 360 cattle, 1,500 sheep and 350 pigs. Basil and his son, James, expanded weekly sales to some 500 prime cattle, 2,000 sheep and 400 pigs until, in 1985, James retired. Soon afterwards, the mart directors sold the Boroughbridge Farmers Auction Company to the Thirsk Farmers Mart for, reputedly, somewhere around half a million pounds, but they found trading difficult and closed the mart in 1990.

Hornby Cranswick commenced weekly sales in **Bridlington** in 1900. He had held sales in the area through the latter half of the nineteenth century and by 1920 the family was trading as Cranswick and Cranswick. The local council erected a new mart in 1920, between the old and new towns, and Cranswick`s carried on a useful weekly fatstock business there until 1970, by which time the senior auctioneer was John Maxwell. Cranswick`s had attended Driffield Mart, some ten miles away, for many years and, in 1980, they joined the consortium of auctioneers there with their auctioneer, Peter Moore, selling the ewes. He then carried on the land and estate agency business in Bridlington which continues to the present day.

Sitting astride the A1, **Doncaster**, famed for its horse racing and horse sales was an obvious location for a successful livestock market. T.R. Miller was holding sales there in 1870 and Thomas Shearman opened his auction sales of sheep at the Saturday market in around 1880, including special sales of rams. Later he opened his own auction mart for fatstock on Tuesdays.

After the Great War, five other auctioneers sold every week in Doncaster Cattle Market- Pennington and Taylor, F.G. Johnstone with Shearmans, who joined George Hawson & Sons, R Jennings & Son and John Dawson & Sons. In one week in September 1930 they reported a total of 352 fat cattle, 702 fat sheep and 30 fat pigs on Tuesday and a good show of Irish cattle with a few dairy cattle and sheep on Saturday. On a week in 1960, the entry was 150 fat cattle, 260 sheep and 280 pigs. On the store stock day 1,065 pigs were sold and 210 store cattle made up to £74. Two auctioneers balloted to sell the fatstock on Tuesday but on Saturday it was more by habit when Shearmans usually sold sheep at 12.30 and Johnstone at 2.30. Johnstone would also sell some twenty or thirty dairy cattle on a Wednesday. All the firms selling at Doncaster joined together in 1970 as Doncaster Market Auctions. The closure, in due course, of the city abattoir had a major effect on the mart, which lost several important buyers and closed in 2001. A member of the Shearman family, Oliver, is a livestock auctioneer in Scotland.

Driffield Cattle Market

In 1880 Francis Cook, a Land Agent, started his auction sales of cattle, sheep and pigs in **Driffield** where, as his business grew, he took Eric Dee into partnership. Driffield is the principal town of the Yorkshire Wolds and sales there had first been held in around 1860 by George Clark. By about 1900, five firms were operating there and, about this time, Lord Downe sold his historic rights to hold a market in Driffield to this syndicate but William Cooper, of Boulton and Cooper of Malton, swooped and bought the Driffield Cattle Market from Messrs Jarrett. This caused quite a stir which only died down once Cooper sold the market on to the auctioneers who used it. By 1935 six auctioneers were involved - Cook and Dee, F.W. Ullyott & Sons, Cranswick and Cranswick, Megginson, Shipley & Wharram. They initially shared the municipal market every Tuesday and later agreed to change the market day to Thursday, thus creating a very busy centre selling 500 fat cattle and 1,500 sheep most weeks of the year with a few days when a thousand cattle were sold. Part of the market was owned by the auctioneers who modernised and improved it to accommodate a thriving business, including a large store cattle trade. The part owned by the town council is now a car park. Mr Dee took Peter Stephen Atkinson onto his staff in 1934 and it became Dee & Atkinson, then it became Dee, Atkinson & Harrison, who carried on a large throughput until 2002 when the local council suggested that the mart might move out of town and the auctioneers got a good offer for their site. Dee and Atkinson and Ullyotts also sold weekly at Beverley and Hull and continue as land and estate agents, but not as auctioneers. The poultry mart in Driffield was operated by F Purdom and Son who sold everything – from large numbers of geese[34] right down to brambles, at 6d per lb and mushrooms at the same price. The auctioneer deducted one penny in the shilling (a twelfth) for his commission.

Boroughbridge Cattle Market
(North Yorkshire County Archives)

In 1890 the town council of **Goole** opened a small market where three local auctioneers, Mr Thorpe, Mr Clegg and Mr J.B. Johnston, advertised fatstock sales as Goole Farmers and Butchers; however by 1914 only Mr Clegg operated and his sons continued to average some 150 fat cattle and 400 fat pigs every week for many years. In 1980

34 An acre of good stubble would feed 15 geese for the Christmas market.

they bought the freehold of the mart, which, by then, was in Bridge Street, beside the docks. Foot and Mouth Disease brought the closure of the mart in 2001 compounded by the fact that the surrounding farms had stopped grazing cattle and now grew arable crops, which were proving more profitable. After the Second World War Clegg`s also sold weekly in Hull and Beverley Marts under the partnership of Clegg and Wright. They continue as land and estate agents in Goole.

About four miles north of Goole lies **Howden** where a Mr Pratt held sales in the Wellington Hotel yard from circa 1870 with other auctioneers subsequently involved. The 1910 Christmas Show at the Wellington Hotel Mart featured 14 show cattle and 50 sheep together with 30 other good cattle and 100 other sheep. Sales at this mart are not recorded after 1918. Horses continued to be sold for many years at Howden Horse Show by Mr Tipling, auctioneer, Tadcaster.

From around 1900 there was a small farmers` mart at **Guisborough**, the Guisborough Auction Company. This was never a very ambitious operation but it traded on for some eighty years until it merged with Stokesley Mart with half the directors of the merged company being from each mart. Three years later, in 1983, Guisborough closed with the business being transferred to Stokesley.

Hellifield Auction Mart

An auctioneer first attended the Fair at **Hawes** in around 1850 and continued to sell some stock in the Cattle Market area off the Main Street until 1887, at which time Mr T.T. Iveson (see Chapter 13) started additional sales there. He steadily expanded the business until 1914 when he moved to concentrate on developing Hexham Mart, leaving Hawes to the Hawes Farmers Auction Mart Company. Under Raymond Lund, assisted by Maurice Scott, Hawes remains one of the major sheep marts in Northern England, renowned for its sales of Mule breeding sheep and Bluefaced Leicester rams, which run on late into the night.

The energetic T.T. Iveson was also involved with **Hellifield** where, in the 1870`s, a group of farmers led by the Iveson family set up an auction mart near the railway station and, with T.T. Iveson as chairman and manager, it flourished. In 1888 the Ivesons built a new mart in Hexham and, thereafter, T.T. carried on as chairman, but day-to-day management at Hellifield was in the hands of the chief auctioneer. Around 1950 Malcolm Skidmore took an interest in the mart and bought any shares available. With boundless energy, he expanded the Hellifield Auction Mart Co Ltd into a leading centre where Laurie Kay sold the fat cattle. On a Monday 300 calved dairy cattle were sold with 200 to 300 in-calf cattle on a Tuesday followed, on Thursday, by some 4,000 fat sheep and 400 fat cattle. In 1986, Craven Cattle Marts of Skipton took over the business and ran it until 1990 when they hoped to transfer the Hellifield business to their new mart at Skipton; this hope was not to be fulfilled as much business was lost to Gisburn.

Holmfirth, on the edge of the Peak District National Park, is best known as the home of the long running T.V, comedy Last of the Summer Wine but it also boasts an auction mart which trades every Tuesday. A.E. Wilby and Son were operating as auctioneers in Holmfirth prior to 1866 at which time William Sykes of William Sykes and Son commenced trading. William Sykes was succeeded by his son Herbert, and then by Herbert`s nephew, William Sykes Shaw. They were joined by Vernon Dixon 1951 but, in that same year, the present firm, Holmfirth Attested Auction Mart, was established when forty four local farmers took shares of £100 each to set up the mart in opposition to the existing mart which did not offer the facilities to sell TB Attested cattle. These £100 shares now trade at up to £33,000 each and many are still held by the original contributing families. Vernon Dixon took over William Sykes and Son in 1959 but they had ceased selling in Holmfirth. Considerable modernisation was carried out at the ongoing Attested Mart in the 1980`s and the mart sells some 20,000 head of stock each year. Horse, pony and tack auctions are held weekly on a Saturday. Vernon Dixon`s son Paul who sold in Penistone and Huddersfield has been involved with sheep sales at Holmfirth since 2008. Holmfirth Mart was recently deservedly described as "one of Yorkshire`s agricultural gems". William Sykes & Son continue on as estate agents and furniture and property auctioneers.

Paul Dixon, auctioneer, Holmfirth

Holmfirth Auction Mart

Twenty four miles from Holmfirth, in the Peak District Park and the county of Derbyshire was the very popular village mart of **Hope**. In its earlier days, it was owned by Siddal and Reynolds of Wadsley Bridge Mart in the northern suburbs of Sheffield. Economy was their watchword and as their printing works neighbour threw out unsold calendars, the auctioneer would cut them up to use as calf numbers. Latterly the mart was carried on by Alastair Sneddon of Bagshaws until Foot and Mouth Disease in 2001. New regulations called for concrete floors and steel pens with the result that, like many other old-fashioned marts, it did not re-open. The loyal supporters were asked to go to Bakewell, where the firm had a very large modern mart, which still flourishes.

In 1718 the cattle market in **Huddersfield** was between Kirkgate and North Gate and tenants of the market were instructed to charge "only the usual rate of tolls as were charged in times past". There must have been some success as, by 1837, the entry had to be widened to avoid injury to the customers. At this time Joseph Kaye rented the pig market and controlled sales there but whether or not he was selling by auction is not known. The cattle and sheep market was later moved to a more spacious site in order to accommodate an auction ring, where it remained for many years with William Sykes & Son conducting the sales. A Sykes had possibly first become involved with the market as the Town Crier who rang his bell and cried the opening and closing of the market for which he was paid a fee of maybe one shilling a day. Latterly Vernon Dixon and, in turn his son, Paul Dixon, owners of William Sykes and Son were auctioneers and there were consistent weekly numbers of maybe one hundred plus prime cattle and several hundred sheep with large numbers of pigs, up to 1,000 head towards Christmas time. In 1996 Huddersfield was closed as the local authority sold the site for re-development and the sales were moved, as far as possible, to Penistone where William Sykes & Son also sold.

Penistone Auction Mart

In 1890 the town council of **Knaresborough** let a few pens in the town centre to a Mr Horner to hold a Monday fatstock sale. In 1900 Joseph Horner sold at his Christmas show 168 cattle, 530 sheep and 310 pigs. Then in 1909 George Thornton provided a few pens in a stable yard and this became the early mart. At a 1935 weekly sale Thornton and Linley sold 125 fat cattle, 600 fat sheep, 60 store cattle and 518 pigs on Monday and 2,200 store sheep on Wednesday 2nd October. Their firm continued trading until 2001 but was unable to re-open after the Foot and Mouth epidemic despite a strong fight by auctioneer, Ian Thornton, who, in the local press branded new government regulations as bureaucracy gone mad. The difficulties of the mart site, which was behind the Borough Bailiff pub, meant that compliance with various requirements was extremely difficult. Despite, or perhaps because of, Mr Thornton stating that the rules had been made by boffins in Whitehall wearing white coats and without any practical knowledge of how things were run (nothing new there then), he got little sympathy with a DEFRA spokesman emphasising that "it is up to auctioneers to decide on commercial grounds whether to undertake the necessary changes" to enable their marts to trade.

Knaresborough, 1906
(North Yorkshire County Archives)

Sales commenced in **Kirkbymoorside**, now on the southern border of the North Yorkshire National Park, in 1887. In due course, this became a Farmers Mart Company with Tom Kendrew as manager and auctioneer; he took David Dennis into partnership in 1954. When David died in 1968 he was selling weekly at Kirkbymoorside where some 70 fat cattle, 450 fat sheep and 50 pigs would appear every Tuesday. He also sold for Cundall`s on Monday at York and on Friday and Saturday at Malton. Cundall`s took over Kendrew and Dennis in 1968, with Mrs Kendrew, a great character, assisting for some years. The mart closed in 1973. Both Paul and Stephen Dennis now work with Turners at Bentham, one in livestock and the other in property. Each only a few miles from Kirkbymoorside are Fadmoor and Wombleton where Cundall`s conduct well supported traditional seasonal sheep sales each year.

Four annual fairs and fortnightly markets were, for many years, held in the historic market town of **Leyburn** in Wensleydale. The earliest auctioneer here was a John E. Atkinson who held sales but probably not on a permanent basis. He certainly held "The Great Annual Sheep Sale" in the paddock adjoining the Bolton Arms Hotel. Atkinson`s business was superseded in 1894 by that of Joseph Hopper whose

Leyburn Livestock Auction Mart, AGM, 1991

Leyburn Manager, Stephen Walker

Leyburn Auctioneer, William Braham, with Raymond Rutherford and Peter Lambert

business, J. R. Hopper, had been established in 1886 and continues today as estate agents and furniture auctioneers. The Hoppers had, at the time, wide ranging business and farming interests in the area. By the time of World War I, Joseph Hopper was advertising as attending Leyburn, Hawes and Richmond Markets with large periodical sales held at Leyburn and Masham during Spring and Autumn. He also sold at Bedale Mart in the 1920`s. Joseph was a participant, and presumably moving spirit, in the 1918 establishment of the Leyburn Livestock Auction Mart Company which operated from the mart`s present Richmond Road site. The share capital was £6,000 in 1,200 £5 shares and with a large board of directors ("not less than ten, nor more than twelve") specified. Present manager, Stephen Walker, has recently written the centenary history of Leyburn Livestock Auction Mart Company and his researches in the minute books confirm that the regular theme throughout the mart`s history has been the collection of overdue debt; a theme unfortunately common to all mart histories. A Mr Wrathall was auctioneer for some years from the 1930`s until his death in 1950. He was certainly a character and, on one occasion, the board requested that the chairman write to tell him that they "disapproved of his drinking whilst in the rostrum"; falling to the temptations of drink being another unfortunate theme with auctioneers over the years. Wrathall was succeeded as principal auctioneer by H. T. Metcalfe of Scotton. Dairy cattle sales were a mainstay of the company at this time and Teeswater sheep a speciality. The mart continues to thrive today. In 2016 the Smithfield Peterborough Fat Stock Show was won by a calf bought at Leyburn in March by Jennifer Hyslop of Girvan which sold for £10,000. Significant numbers of rearing calves pass through the mart and William Bramham has been the auctioneer of these for over thirty years; also selling, in his time, at Barnard Castle and Pateley Bridge. Stephen Walker, the current general manager who worked with the author at Annan Mart for a time, succeeded Derek Alsop who was the previous chief auctioneer of the company. Derek died recently and as they say "when a man dies a library closes". He also played a leading role at Northallerton mart where he was a director. Philip Terry is the energetic present chairman of Leyburn Mart.

Malton Auction Market, an early view

Robert Preston held weekly sales on the Green at **Long Preston** in the Yorkshire Dales from about 1880 onwards. In 1930 his grandson built a new mart where 60 to 80 fat cattle and 600 to 1,000 fat sheep were sold every Monday until his death in 1953. His widow carried on, with David Hill as auctioneer, until 1959 when James Thompson of Kirkby Lonsdale bought the mart and moved the sale day to a Friday. The mart closed in 1972. It had traditionally depended on dealers to provide the fatstock entries and was never a large store stock centre.

Malton Auction Market, the way we were

Malton, approximately midway between York and Scarborough is in a natural location for a successful auction mart. In the mid 19th century John Boulton combined inn-keeping with auctioneering and was also known to be "one of the finest judges of sheep in the three Ridings".

His son Robert, an auctioneer, was a cricketer and sportsman who included curling and coursing in his enthusiasms. For a time, he conducted the yearling horse sales at Doncaster. By 1868 fortnightly sales of livestock were held in Malton and, in 1891 Robert Boulton changed to weekly sales which he successfully developed so that in the year of his death, 1901, over 10,000 fat cattle, 79,000 sheep and 10,000 pigs passed through the mart. The main market day was Saturday with fatstock sold on Tuesdays. Robert Boulton was assisted in his business, firstly by Mr Constable and then, from 1888

Malton Market, 2017
(Photo, Colin Grice)

onwards by William Cooper who commenced as cashier to Boulton and Constable and in due course became a partner in the firm of Boulton and Cooper whose name carries down to this day. As well as developing the trade in Malton, which was on land provided by the Fitzwilliam Estate, the firm built a mart at Seamer Junction and sold in Pickering Mart. William Cooper also farmed in a substantial way and during World War I, in common with many other auctioneers, he did an immense amount of work in organising auction sales to benefit the Red Cross. He also developed his firm`s Vale of Derwent Horse Mart in Castlegate, Malton so that, by early in the 20th century, it was one of the principal centres in North Yorkshire for the sale of horses and carriages. Parallel with Boulton and Cooper, the firm of Joseph Cundall and Sons also sold in Malton market, as did Dee and Atkinson of Driffield for a time. Cundalls first sales there were held in about 1885. They traded right through until the 1980`s when they merged with the surveying firm of Wells of Hull. After 1954 John Cundall had become the driving force in the firm and by 1970 Malton frequently sold a thousand fat cattle in a week with Cundalls leading the way. By 1990 they advertised as Wells Cundall Agriculture with 1,400 suckled calves on one day in the Cattle Market, Malton, and a sale at Farndale in conjunction with Chris Clubley & Co of Pocklington. At this time, Boulton and Cooper were acquired by Stephensons of York and the mart is now operated as

Masham Farmers' Live Stock Mart Co. Ltd.

WEDNESDAY, SEPT. 25th, 1946

SPECIAL SHOW & SALE OF

2,400 HALF-BRED GIMMER SHEARLINGS,

Ewes, D.C., XX and Scotch Lambs,

Newly Calven and In Calf Cows and Heifers,

and all other Classes of Cattle,

Which will be sold

at the ABOVE MART, MASHAM.

All Sheep to be penned ready for Judging at 10-30 a.m.

SALE TO COMMENCE AT 11 O'CLOCK PROMPT.

Vendors are responsible for Sheep until taken over by the Purchaser.

Every assistance will be given to buyers in loading. Any to stay overnight will be kept at owner's risk at reasonable charges.

Also will Vendors give every assistance to Buyers.

H. HEWISON, Secretary,
Market Place, Masham.

TERMS CASH.

Manager—
H. W. VERITY,
Fearby, Masham.

Auctioneer—
D. JOHNSTON,
Green Hammerton, York.

PHONE :—**Green Hammerton 247**

NO refreshments at the Mart.

Metcalfes & Harper, Printers, Ripon.

a working partnership between Cundalls and Boulton and Cooper Stephensons. Cundalls are now owned by the Watson family. The mart at Malton is on land owned by the Fitzwilliam Family Trust who, in 2010, offered a new site at Eden Camp outside the town so that the old town centre site could be re-developed. The Malton and Ryedale Farmers Mart Ltd was set up but for eight years there has been no real progress and Malton remains a traditional town centre market with pens for sheep seasonally erected in the streets.

Pigot`s Commercial Directory for 1829 tells us that **Market Weighton**, in the East Riding, was host to fairs on the 14th of May and the 25th of September and that "the latter, at one period, was the largest sheep fair in England and still remains very considerable". Previously, in 1812, it was reported that there were "frequently 70,000 to 80,000 fat or nearly fat sheep upon the ground". From about 1880 to 1914, frequent sales by auction of store stock were held in the Londesborough Hotel Yard, latterly by R.M. English and Son but did not continue after World War I.

George Chapman's prize pen at Masham sheep sale, 1971. Matt Pallister auctioneer

Masham was historically the site of very important sheep fairs where, in 1887, Robert Imeson commenced holding auction sales. These developed into fortnightly Wednesday sales of sheep, dairy cattle and store cattle. After the First World War a Farmers Auction Company, with a capital of £3,000, took over but profits were never large, in spite of charging a minimum commission of four old pence per sheep allied to an overall rate of four pence per £1. After the Second World War they employed Don Johnston and Matt Pallister as auctioneers. Latterly Willis Graham, a local farmer, was mart manager and in 1998 Craven Cattle Marts from Skipton conducted sales at the mart and held weekly evening sheep sales until, by 2006, the pressure was on and, despite the efforts of Willis Graham and others, the board decided to close the mart and it never re-opened. Perhaps the shareholders, there were only 3,000 shares, were tempted by the high value of the site when compared with many years of receiving no dividends. Masham was surrounded by other marts of which Leyburn continues and Bedale and Ripon have closed. However the local supporters did not give in when the mart closed and restarted the historic Great Fair in the town`s large market square in the last week in September, where sheep can be seen standing in the streets in pens as they have done for a thousand years. By 2006 some parts of the mart premises were being sold off and in 2017 Messrs Lister Haigh were instructed to sell the remaining two and half acre site with a target price of a million pounds.

Masham Mart, the finale, 16th February 2008, sale of pens etc.

Northallerton was and remains an important market centre with, at one time, three auction marts. Mr Atkinson, at the Station Mart and Mr Malpas, at another site, both commenced business in around 1875. However it was the Applegarth Mart in the town centre just off High Street and across the road from Malpas Mart, that was opened in 1887 by Mr Todd as John Todd`s Northallerton Store Stock Market, that continues to thrive today, taking over the Malpas Mart in 1954 and the Station Mart in 1975 and trading for over a hundred years as Northallerton Auctions Ltd. In 1880 Atkinson reported 130 prime cattle, 680 sheep and 210 pigs at his weekly sale in November. In 1943, Malpas were temporarily put out of business for a few months by a German bomb. The previous year, in 1942 Derek Allsop had started as an auctioneer at Applegarth mart, ten years later becoming a director and then chairman of a very successful auction business selling around 500 cattle and 2,000 sheep each week. Surprisingly, he was also senior auctioneer on a freelance basis at Leyburn mart, less than twenty miles away, where his integrity was never questioned. The decade 1990 to 1999 was a poor period with turnover falling from £26 million to £12 million and stock numbers down to 35,000 store and 35,000 fat sheep, 14,000 fat and 9,000 store cattle and only 8,000 pigs by 2000. Mr R.E. Phillips, the chairman, successfully overcame the dissatisfaction of many shareholders and the mart is now going well. After 2001 Northallerton continued to sell some 200 fat cattle and 1,500 fat sheep each week under chief auctioneer, Giles Drew from Worcester. Weekly sales are held on Tuesday and Wednesday. There is also a busy estate agency and furniture department.

If readers fear that **Otley** has been omitted from this chapter, then they are referred to Chapter 13 and to the section on the Lister family.

Situated north of Harrogate, at **Pateley Bridge** in Nidderdale is a mart where pre-1939 the sales were organised by Mr Lister, who reported the sale of 5,100 store sheep on 1st October 1935. This mart had first traded, from around 1900, as Pateley Bridge Cattle Market and in 1951 it was taken over by Atkinson and Worthey who further developed the sales of cattle and from 1953 ran it jointly with **Sawley** Mart, about eight miles distant, under the name of Sawley and District Farmers Auction Company, with sixty five shareholders. In 2010 the name was changed to Pateley Bridge and District Farmers Auction Company with David and Judy Middlemiss as the energetic managers and directors and the popular William Bramham as chief auctioneer; they hold fortnightly sales of cattle, with 150 at their

Pateley Bridge Auction Mart

Pateley Bridge, with Matthew Probert in the box

February show, and their weekly sheep sales average around 1,000. There had, in fact, been quite a battle to keep things going after the

Foot and Mouth outbreak and it was only the perseverance of one or two directors and a number of shareholders, who helped out without charge, that the mart is still successfully operating today.

Lowther & Son were trading as auctioneers in the historic town of **Pontefract** from 1863 with a Mr Dodds also holding sales from about 1870. Bentley & Sons were in business from around 1895 and would sell, on a Tuesday in 1900, some 50 fat beasts and 150 fat sheep. They sold pigs also but the pig market was closed for a time in that year due to a disease outbreak. Numbers would double for their Christmas Show. Bentley`s were succeeded by Sykes and Son, taken over by Taylor and Co. in 1939. Bendall & Co. were also selling in Pontefract by 1954 and we can note that, on a specific market day in 1960, Taylors were selling 27 fat cattle, 51 sheep and 72 pigs whilst Bendall`s had 45 cattle, 30 sheep and 150 pigs. These numbers were modest and the mart closed in about 1965.

There was not a mart at **Reeth** but it is given here as an example of the many seasonal sheep sales, with temporary penning, which were held throughout the North of England[35]. Deep in Swaledale, Reeth obtained a Royal Charter to hold the Bartle, or St Bartholomew`s Fair from William III and Queen Mary in the late 17th century. Such was its success that, at the height of local prosperity, six fairs were held each year. An auctioneer first appeared to sell sheep at the fair in around 1860 and by the turn of the century two sheep sales were held each autumn in a field opposite the pub at the bottom of the village, later owned by the Browns of Swale Hall. The sheep were driven in from a radius of around ten miles to be sold by three generations of the Addison Family, Tom, Joe and Trevor, who would sell 3,000 to 4,000 on the day in two rings, a lamb ring and a ewe ring, where the very best of Swaledale sheep were offered. By the 1950`s sales were held in conjunction with Leyburn Auction Mart and the sale venue was moved on several occasions. From the 1960s there were regular discussions as to whether the sale should be moved to the mart. The sellers were against this. Some controversy must have arisen in 1967 for Mr F.H. Brown the landlord of the sale site stated that Leyburn Mart would not be allowed to participate in the sheep sales which were held on his land. However a new site was found at Windy Hall but the quarrel must have been patched up and the sale returned to Mr Brown`s Reeth Showfield in the following year. From then on, the first sale was to be conducted by Addisons and Leyburn Mart and

---

35 See also Chapter 10

the second sale by Addisons and Mr Brown. However, as time went on, lorries and trailers tempted farmers to go to marts with covered sale rings and more comfortable accommodation and, in 1991, the last temporary hurdle pens were erected and the last village sale at Reeth was held.

In **Richmond, s**ales were held in the large Square in front of the Bishop's Hotel until 1920 when the Earl of Zetland provided land for a small auction mart which held fortnightly fat sales as well as store sales. Three generations of Addisons wielded the gavel until 1987 when it joined with Stokesley Auction Co who, in the early 1990s, closed the Richmond mart and let the Stokesley mart to the Acklington Auction Mart Co. with Stephen Walker and George Potts acting as auctioneers and also conducting the sales at Bishop Auckland.

Ruswarp Auction Mart, 2011
(Photo, Colin Grice)

Ripon Farmers Mart traded from 1890 until 1935 when the Kirby family took over. In 1957 at their weekly Monday sale at the end of March they were getting 110 pork and store pigs, 500 fat sheep and 130 fat beasts, which topped at £24. Soon afterwards, Mr R.H.

James became managing auctioneer, succeeding Tom Kirkby who had abolished the £1 maximum charge for selling fat cattle. By 1967 the corresponding sale had 195 fat cattle and 710 sheep but only 31

Ripon Auction Mart, 2005

In 1887 Charles Smith set up in business in **Whitby** as an auctioneer and estate agent. He was in due course, succeeded by his grandson, also Charles Smith who took a great Yorkshire cricketer Allan Richardson as his partner. Allan ran a small market at **Lythe** a few miles north of Whitby which they closed to enable them to concentrate on the mart at **Ruswarp** which they bought and improved in around 1920, trading as Richardson and Smith. Prior to that, since 1902, Ruswarp had been operated by the Whitby and District Farmers and Butchers Auction Company. Richardson and Smith continues to this day with members of the Smith family still partners and Robert Smith together with the Halley family, currently Ian Halley assisted by his daughter, Danielle, as the cattle auctioneers. They handle up to 800 store cattle at their major autumn sales. They sell on two days a week with good store sales on a Wednesday, weekly sales of fat sheep on a Monday and small fortnightly sales of fat cattle, at their popular traditional mart with the steam trains of the Moors Railway still passing. The firm used to hold seasonal sheep sales at Glaisdale, some 10 miles away, and also at Castleton; these sales ceased in 2001, after Foot and Mouth Disease, and the sheep from there are now sold at Ruswarp. Today the firm still has a busy estate and land agency practice.

Rotherham Cattle Market, 1905

Ripon Auction Mart, all that remains

Three fairs a year were traditionally held in **Selby** for the sale of "horses, horned cattle and sheep". Today, the town remains a busy and successful marketing centre boasting the largest weekly prime cattle and pig sales in Yorkshire with some 400 to 500 cattle, 1,000 plus sheep and 400 to 500 pigs on many weeks. The mart was originally in James Street but moved to new modern premises in 1989. Although other auctioneers have sold in the mart over the years, the Bartle family have been the principal operators since 1873, and remain so today. Brian Bartle is the present chairman of the mart company with his cousin, Philip, the company secretary. Chris Clubley is a regular auctioneer and shares the selling with general manager Richard Haigh who trained in Scotland and came to Selby in the mid 1990s to replace the very popular Robin Screeton who did so much to make the mart the success that it remains. The Bartles have also, has his own multi-branch property business in the area.

Ruswarp Mart, Christmas 1965

Markets were held in **Sheffield** from at least the 13th century and all the market rights belonged to the Lord of the Manor. The Lord of the Manor in the late 18th century was the Earl of Surrey, later Duke of Norfolk and he was petitioned by the local inhabitants, tired of the increasing numbers of animals being driven through the streets, to provide them with a larger and more accessible market place. Thus, in 1786, the cattle market was relocated from Bullstake to The

Wicker and new slaughter houses were provided. In 1827 the then Duke of Norfolk was empowered by an Act of Parliament to provide a new livestock market at Smithfield beside the river Don. In 1886 auction rings were provided, either on the same site or nearby, to accommodate the new method of selling. However the sales were never very large and one commentator reported that there were more tryers than buyers, although in September large numbers of Irish geese would be sold to farmers at two or three shillings each to feed on the stubbles for Christmas. Latterly sales were held at Wadsley Bridge, a suburb of Sheffield, about three miles from the city centre, but this mart closed in the early 1960s.

Skipton Old Auction Mart

Skeap Tun can be translated from Old English into sheep-town and thus **Skipton** which still justifies its name today with a successful auction mart selling some third of a million sheep each year, not to mention 20,000 or so cattle. With market charters dating back to the reign of King John, Skipton was home to large fortnightly livestock markets in High Street said, in 1838, to be "inferior only to the show at Smithfield; the sums of money which change hands every fortnight are enormous". A great proportion of the cattle sold were Irish stores. The sheer scale of the market resulted latterly in much local opposition including from the Court Leet in 1809 who complained that the sheep pens placed on the fronts on the Monday evenings by the local publicans[36] were a nuisance and later criticism from the Craven Herald in 1875 commented on "the intolerable nuisance caused by the gathering of so many cattle in a space sadly too limited for their requirements". Finally, in 1906, in response to pressure from shopkeepers who complained that customers couldn`t get to their premises on market days, the market was relocated to a site behind the town hall, in Jerry Croft. By 1880, several auctioneers were practicing their trade at this market and, in 1886, John Throup and W.H. Davis established a small mart with an initial offering of eighty-one fat sheep, twelve cattle and two carts. The Skipton Auction Mart was established on Broughton Road in about 1894 and the present incumbents, Craven Cattle Marts were incorporated as a limited company in 1921. For many years the mart was under the management of Matt Pallister and Albert Jackson. In 1935 they claimed that their regular sales of 500 in-calf cows were the largest in the Kingdom; at about the same time they reported 185 fat cattle, 1,600 fat sheep and 400 pigs at their weekly prime stock sale. By the 1980s the congestion pressure on the town centre site had become very great and a particularly fine new mart was erected a mile out of town under the leadership of chairman, Stephen Butcher and mart secretary Paddy Wrightson. It was opened in 1990 and boasts a characterful traditional façade concealing the modern workings of the mart including a sale ring which converts into a theatre. Over the years there has been substantial further investment in the premises and recent extensive scheme of improvement has just been completed. It is now a leading centre for fat and store stock with many pedigree and special sales under the direction Jeremy Eaton and Ted Ogden. Their sales of working sheepdogs are renowned.

The residents of **Stokesley** in 1823 included two auctioneers but it is not recorded whether they sold livestock. Fairs for the sale of cattle were held on the Saturdays preceding Palm Sunday and Trinity Sunday. There were two marts in the town from the end of the 19th century, the Station Mart and the Town Mart. In 1900, William Hedley at the Town Mart was selling 61 cattle, 658 sheep and 219 pigs on one day; ten years later, at the Station Mart in March 1910,

36 For instance, the Black Horse Inn would erect forty sheep pens and let them out at sixpence per pen.

Richardson and Dobson had 150 pigs, 280 fat sheep, 30 store cattle, 32 dairy cows (at up to £21-10/-) and 90 fat cattle (making up to £23). The two marts amalgamated on one site in 1965 and changes in operators concluded with the mart being run by Penrith Farmers & Kidds of Cumbria, Stephen Aitken, auctioneer, from 1995 until its closure in 2004.

In 1869 Matthew Thomlinson set up his board in a yard in the main street in **Tadcaster** and attracted so much stock to sell that Jacob Stephenson was tempted to set up in opposition. He attracted a lot of custom, which encouraged him to send his son, Richard, to attend York market and in 1900 the Stephensons moved all their business to York, leaving Thomlinson and his son, Edmund to sell by auction in Tadcaster every week. In 1935, they were averaging 180 fat cattle, 450 fat sheep, 300 pigs and 30 milk cows. Thomlinsons also sold weekly at Wetherby and York with Reg James doing much of the selling, and by 1960 their firm was also selling in Wakefield Market. Tadcaster closed in 1990.

Established in around 1875, in 1880 George Freeman's Mart in **Thirsk** were advertising a Christmas Sale of 130 cattle, 320 sheep, 300 pigs and 15 dairy cattle. In 1906 a farmers group attempted to take over Freeman`s but were rebuffed so, in the following year, they raised £2,500 of capital to build a new mart in Station Road on land provided by the Turton family with Seth Kirby as auctioneer. Freeman`s business closed towards the end of World War I and the Farmers Mart went from strength to strength. For a short period, 1965 to 1972, the directors leased the mart to Seth Kirby & Sons before taking it back under their control. It became a large fatstock market, with up to 1,000 prime cattle at its peak and from 1972 massive improvements were made to the buildings and store stock sales were expanded. In 2006 the mart moved out to a new Thirsk Rural Centre with a member of the Kirby family as company secretary and Tony Thomson in the rostrum. Today, Andrew Roberts is the manager and Emma Coupland sells the sheep. For many years the Woodhead family have been involved, owning over a third of the shares, and today Ian Woodhead is chairman of the company, which also owns Joplings Estate Agents. Ian`s father Harry was very much responsible for the ongoing survival and success of the mart and proposed, at one time, that Thirsk, Ripon and Northallerton marts should merge in one new mart. This far-sighted scheme came to nothing.

In 1240, the Knights Templar were granted a charter to hold a market in **Wetherby.** In due course livestock trading would take place in Victoria Street and High Street. As far as auctioneers were concerned, Sergeant and Son were in business in 1875 and, by 1880, Matthew Thomlinson had established substantial weekly sales of fatstock on Mondays and some large seasonal sales of store sheep and cattle. By 1900 he had been joined by his sons, Edmund and William, with weekly sales of 80 prime cattle, 200 prime sheep and 100 pigs, numbers that were doubled by 1960 and Thomlinson and Sons proved to be a flourishing company. Back in 1912, old cottages in the Horsefair area of the town centre had been demolished and a new auction mart created. Thomlinsons finally closed this mart in 2001 after the Foot and Mouth Disease outbreak, which finally knocked the spirit out of so many old firms. Marks and Spencers stands on part of the site. Thomlinsons still trade as estate agents in the town.

CHAPTER XVI
1940 – 1954, THE LONG CONTROL AND AFTERWARDS

Three pre-war years of meticulous planning by civil servants and auctioneers were repaid by a smooth conversion to meat rationing in the six months after the Second World War was declared. The stock was graded at the mart, allocated to the appropriate abattoir and then delivered to the butchers` shops. The butcher would then sell a weekly ration book allowance of meat to each customer. The weekly value of the allowance fluctuated between one shilling and threepence and two shillings depending on the available national supply. This system worked smoothly and persisted for some fourteen years and with the consequent rise in prices for fatstock, all purchased by the Ministry of Food, farmers entered a long period of prosperity.

Selby Auction Mart

For auction marts with a reasonable throughput of store stock, it was initially a tranquil period with no bad debts and no canvassing for fat stock as all farms were allocated to their local auctioneers in each town, the work was divided among them and went amicably on for fourteen years.

Leyburn Spring Bull Show, 1951

Tranquillity soon came to an end as the Royal Navy rescued the Army at Dunkirk, auction staffs, along with their fellow citizens, joined the Home Guard, enlisted as Air Raid Wardens or enrolled as Special Constables. Many marts had some fields, which, for the first time since the First World War, grew crops and the auction mart men learned how to stook the corn and stack it.

Entries of livestock had to be made a few days before the weekly grading took place so that the local auctioneer-grader could receive his allocation instructions based on what a railway waggon would hold. The last on the list of destinations was a large wholesale slaughterhouse, which could deal with any shortfall or excess. For some strange reason among the allocations from my home mart of Annan in Dumfriesshire a waggon load of sheep went every week to Bradford.

Skipton Beef Shorthorn Sale, the author's son and grandson in the ring

During the war leading auctioneers gave a great deal of thought to the future of the marts and found great support among both sellers and buyers, who believed, as most still do to-day, that it is the best way to get the true market value. To achieve this, marts and their staff needed to be modernised with better scholastic training in many aspects of their work.

Back in 1919, a serious shortage of trained estate agents and surveyors, was met by establishing a College of Estate Management in London. In due course, this received a Royal Charter and later moved to a new campus in Reading, now part of the university.

They had a large and efficient department that provided postal tuition for the examinations of the Surveyors, the Auctioneers and Estate Agents, and the Land Agents societies. However, after seventy years, they rather lost interest in the training of livestock auctioneers, tuition and exams for these are now very successfully masterminded by the Harper Adams University in Shropshire.

In 1949 The Institute of Auctioneers and Estate Agents, Lincolns Inn, London were granted a Royal Charter and qualified auctioneers became "Chartered Auctioneers". In the same year, the Benevolent Fund of the Institute for members and widows improved their support for members and widows. For single persons whose total income in cash and kind from all sources was less than £3 per week their income was brought up to £3 and the minimum for two people living together was increased to five guineas and where home help was needed the income level was increased to £7.50. There were a total of 38 beneficiaries in 1948.

Another cause for celebration was that an annual government licence fee of £10 upon Appraisers and Auctioneers was finally abolished on 5th July 1949 after 104 years. Only Section 7 of the Auctioneers Act of 1845 remained, which required an auctioneer, before commencing any sale, to suspend a ticket or board containing his full name and address at the sale site.

A long-standing custom in Lincolnshire was for livestock auctioneers to charge a part of their commission to buyers and the Institute considered it an appropriate time to try to abolish this impost on purchasers. They did not imagine that within fifty years the buyers commission would become the salvation of all the furniture and fine art auction rooms in Britain.

Among other changes at this time the Stamp Duty Act of 1891 had placed a small stamp duty on valuations, which brought little income to the government and its abolition in 1949 was welcomed by all valuers.

The Annual Provincial Meeting of the Institute was held in 1948, over three days, at the Grand Hotel, Harrogate. To avoid worry to the country folk of Yorkshire and Lancashire, the President sent a letter to all members saying, and I quote, "Although full evening dress is the usual wear for evening functions at the Annual Provincial Meetings, members who have only dinner jackets will be equally welcome". In a few years "white tie and tails" would be a thing of the past.

It was decided to set up an Institute Development Fund as they became a Chartered body and in 1948 some £30,000 was raised by the District Associations. Yorkshire led the way with £3,054 and Lancashire, with the two districts of Manchester and Liverpool (which included Cheshire) contributed £3,980.

A one-day conference of Agricultural members was held under the chairmanship of H.R. Douglas Harding, when it was reported that a new Agricultural Syllabus for the examinations had been adopted, which would include the 1947 Agriculture Act, which had brought in many changes on different aspects of farming and marketing.

Photograph by "The Preston Guardian."

Left to Right: Messrs. Jos. Hothersall, J. Martland, R. T. Handley (Area Livestock Supervisor), John Hothersall (County Chairman, Lancashire N.), R. Clarkson (Chairman, Pilling Auction Mart), R. S. Harrison, J. H. Greenfield (District Chairman, Kendal Collecting Centre), Capt. A. L. Kidd (County Chairman, Westmorland), who made the presentation, Mr. John Hothersall, Junr. (District Chairman, Preston Collecting Centre), and Capt. H. L. Trotter (Deputy Area Livestock Supervisor).

R S Harrison presentation

From 1938 there had been an auctioneers` committee of the Institute and this committee effectively represented the marts in negotiations with the Government and other national bodies. This committee under the chairmanship of Major Richard Woolley from Salisbury worked with my father, Matthew Thomson from the Scottish Institute of Auctioneers, in the pre-war period when a war time meat rationing scheme was being planned. When war came the Ministry of Food requisitioned several large hotels in Colwyn Bay and that became their base. Stanley Harrison of Carlisle worked there throughout the war as Area Livestock Supervisor covering an area from Derby to Carnaervon and north to the Scottish Borders where my father took over as Sub Area Chairman of Auctioneers for the four south west counties in Scotland with responsibility to pay for all stock purchased by the Ministry.

After almost three years of devoted service Stanley had to relinquish his job owing to the death of his brother, which necessitated his return to the family business of Harrison & Hetherington, Carlisle, who, then, had eleven marts, six in England and five in Scotland.

At a presentation dinner, Mr Handley, who succeeded Mr Harrison, said his admiration for Mr Harrison`s work at the Ministry of Food was beyond words. Mr Hothersall of Preston had organised the presentation, which was made by Mr Kidd of Penrith and included a fine sporting gun. I know that the gun would be well used as, on several occasions, as a school boy, I enjoyed beating on Mr Harrison's shoot in Cumberland. The cash left over was donated by Stanley Harrison to the Auctioneers Benevolent Fund.

The auctioneers, from a sense of patriotism, agreed to operate the wartime Collecting Centres for a very modest fee. This was not revised until 1943, when more appropriate payments were received. Cattle 30p, sheep 4p, calves 5p and pigs 7.5p per head. Another review took place in 1946, following regular pressure from the auctioneers` professional bodies.

In 1954 the auctioneers` committee was put on a more formal basis by the creation of the Livestock Auctioneers Markets Committee for England and Wales. This consisted of fifteen Chartered members, four members appointed by the Incorporated Society of Valuers and Auctioneers, four co-opted to represent marts not connected to the two official bodies plus the President of the Scottish Institute of Auctioneers and Appraisers. Representing Scotland for two years, I was impressed by the dedication and practicality of all the committee members. Dick Harrison from Carlisle and Richard Turner from Bentham were from the North West and Yorkshire was represented by Reg Stephenson of York and Eric Dee of Driffield. All of these were particularly valuable contributors to the Committee's affairs. The secretary of the Committee for many years was Mr J. Muir Watt, the value of whose work cannot be overstated.

In 1952 the N.F.U. asked the government to set up a producer-controlled monopoly meat marketing board. A government committee chaired by Lord Lucas had broadly supported such a move in a 1948 report. However, the livestock auctioneers of England, Scotland and Wales supported by some heavyweights among the Conservative members of parliament produced a scheme which combined the free auction system with a deficiency payment that gave a guaranteed price to the farmer.

Details of the scheme are in Appendix 3. The government of the time fortunately believed in a free market and put this into effect with an Annual Price Review coming into operation in 1954. The marts were able to operate freely again and handled very significant numbers of stock throughout the 1950s and 1960s.

Primestock auction sales re-started in 1954 and, to the surprise of the N.F.U. and those who had not wanted to see these auctions continue, every sale in 600 marts went smoothly with sales at satisfying prices and the auctioneers, doubtless, breathing a sigh of relief.

Pre-war patterns changed a little and there were some surprises; for example increased numbers came to some marts such as Bingley, where 400 fat cattle could be sold, many of them fresh from Ireland and, on one day, an unexpected entry of 800 fat pigs caused mayhem. It was the beginning of thirty years of fairly stable prosperity until the time when CJD disease was blamed on cattle and the public were deterred from eating meat with catastrophic results for the industry.

The following table from 1960, covering Mondays and Tuesdays only, demonstrates the strength of the marts at this time.

Yorkshire Livestock Auctioneers Association **YORKSHIRE FAT STOCK MARKETS 1960** Entries of FAT STOCK at the FOLLOWING MARKETS - Monday and Tuesday					
			CATTLE	SHEEP	PIGS
Monday	September 5	BARNSLEY	98	144	148
Monday	September 5	BEVERLEY	49	205	90
Monday	September 5	BOROUGHBRIDGE	216	497	164
Monday	September 5	BRIDLINGTON	18	152	65
Monday	September 5	GOOLE	61	100	70
Monday	September 5	HUDDERSFIELD	96	105	165
Monday	September 5	HULL	256	1,320	590
Monday	September 5	KNARESBOROUGH	64	307	70
Monday	September 5	OTLEY (Wharfedale)	126	660	110
Monday	September 5	OTLEY (Bridge End, Attested)	59	496	---
Monday	September 5	ROTHERHAM	28	34	120
Monday	September 5	SEAMER JUNCTION	133	632	259
Monday	September 5	SKIPTON	136	1,250	10
Monday	September 5	THIRSK	20	183	29
Monday	September 5	WETHERBY	212	493	183

Monday	September 5	WHITKIRK MARKET	40	110	15
Monday	September 5	YORK	170	558	218
Tuesday	September 6	DONCASTER	164	237	264
Tuesday	September 6	DRIFFIELD	66	407	81
Tuesday	September 6	KIRKBY MOORSIDE	24	114	---
Tuesday	September 6	PONTEFRACT	30	54	85
Tuesday	September 6	WAKEFIELD	16	102	92
Tuesday	September 6	PANNAL (Harrogate)	147	580	156

As competition for the marts and encouraged by the ever-present theorists, the N.F.U. were instrumental in the establishment of a Fatstock Marketing Corporation whose aim was to do the whole job and eliminate the marts. Quite the reverse happened, the farmers showed that they preferred freedom of choice.

The Fatstock Marketing Corporation had bought several old established companies of meat wholesalers and bacon curers with a network of slaughterhouses and, for a short time after 1954, they were keen competition. However their performance often did not measure up to their promises and farmers preferred the transparency of the auction system. The only way for the FMC to get prime cattle and sheep was to buy them in the mart and for a while they turned out to be important customers. In the pig section they proved to be more successful but overall they steadily accrued financial losses with the inevitable result that, after about ten years, they went out of business and troubled auctioneers no more.

The sheep trade developed from a situation where Britain imported vast quantities of New Zealand lamb to one where there was a dependence on exports, mostly live, first to France and then to Spain and further along the Mediterranean coast. The animal welfare lobby objected to those journeys for our sheep and despite everyone's efforts, including those of the Farmers Ferry Group, live exports were replaced by deadweight containers of lamb. Such exports were opposed by some French farmers who took the extreme measure of throwing the carcasses into the sea. British farmers and auctioneers retaliated by not buying French wines – but only for a very short period!

These sheep meat exports have now grown to 103,000 tonnes in the last year, of which 85,000 tonnes go to Europe and 18,000 tonnes travel further with some 13,000 tonnes of this ending up in Hong Kong and China. Ewe numbers have fallen by three million in the United Kingdom but the auction marts have maintained their important position in the prime sheep trade.

TESTIMONIAL

presented to

R. S. HARRISON, Esq., F.A.I.

by the

COUNTY AND DISTRICT CHAIRMEN

in appreciation of his Valuable Services

as

Area Livestock Supervisor

for the

NORTH-WESTERN AREA

1939 to 1942

The assertions of death and injury to lambs being exported alive were not borne out by my own experience of sending store lambs to France for several years. We had no reports of death or injury coming back to us and my experience of large numbers of lambs travelling to us from the north of Scotland is, again, that the animals travelled well and trot off the lorry on arrival and immediately settle and graze. The welfare lobbyists forget or ignore the fact that the care of the animals is in the farmer and auctioneer's best interest, both emotionally and financially.

Bad debts were an ongoing problem for many auctioneers as the buyers wanted two or three weeks credit whilst the marts prided themselves on prompt and, more importantly, guaranteed payment to the sellers. The coming of supermarkets made matters much worse as they became such large customers that the wholesale butchers became subservient to them and they, in turn, leant heavily on the marts to help them to support the unreasonable demands of the supermarkets.

In 1970 it was noted that over sixty per cent of fat sheep, fifty per cent of fat cattle and fifteen per cent of fat pigs were still sold in the marts. However marts kept closing and in 1997 the chairman of the Markets Committee, David Thomlinson from Carlisle, would report that the number of marts in England and Wales had fallen below 200, the most recent closures at that time being Beccles and Wickham in East Anglia and Sturminster Newton in Dorset.

The sale of fat pigs fell away particularly steeply. In 1963 21% were sold by auction but by 1990 only 8% went through the marts and numbers continued to fall thereafter.

CHAPTER XVII
THE MARTS OF CHESHIRE

In my search for cattle markets in Yorkshire and Lancashire, inevitably, I often found reference to marts in the neighbouring fertile county of Cheshire, which covers over a thousand square miles with a forty-mile boundary with Lancashire from the hills of the Peak District to Flintshire. I thought that it would be useful to set down what I have discovered, along the way, about these marts[37]. and keep alive a little of their story. Having stood (now sit), in the rostrum for the best part of eighty years I may claim to be reasonably qualified for the job, but accept that Cheshire is a long way out of my own district and that I still have much to learn. I hope that readers will write to me to correct and augment what briefly follows.

Booth Hewitt of Alpraham was a pioneer livestock auctioneer and go-ahead fellow. In 1872 he commenced sales at **Beeston Castle** and developed a series of other livestock sales following the lines of the railways and adjoining the stations at Malpas, Tattenhall Road, Calveley and Worleston. Beeston proved to be his best bet as its advantage over the other sites was that a clause in the agreement between Lord Tollemache, the local estate owner, and the railway company obliged all trains to stop at Beeston. Hewitt was not unlike that other great early auctioneer, Robert Donkin of Northumberland, in that his advertisements and publicity material were expansive, enthusiastic and beautifully written. His mart in Beeston was on a site near the Beeston Castle Hotel and was advertised as Beeston Castle Station Cattle Mart, later to be known as Beeston Castle Smithfield It proved to be successful but, sadly, Booth Hewitt died in 1896, aged only fifty-four.

For some years Booth Hewitt had employed Joseph Wright, a strong Methodist, as a clerk and it was Wright who stepped in at short notice as an auctioneer and who bought the business whilst employing one of Hewitt`s sons, Richard, who was too young to take things over on his own account. The new business traded as Messrs Joseph Wright, also of Alpraham, and flourished. Joseph Wright was supported by

37 The author has been greatly helped by a useful and entertaining small book, "The Fall of the Hammer", including a history of Wright Manley, by Brian Pennington, which was published in 2002 by Cheshire Country Publishing. Particular thanks must also go to Gwyn Williams of Wright Marshall, John Wilson of Whittaker & Biggs and John Brereton of Jones, Peckover & Co.

his two sons, Harold and Joseph W. Wright and they went into a partnership with him in the early 1920s. Harold Wright and Richard Hewitt were the principal auctioneers. The next generation, Alan Wright and another Harold Wright would follow but there was a setback when the older Harold died untimely in 1936. Joseph, senior, died in 1942 and his son, young Joe in 1952.

Over the years progressive improvements and extensions were made to the Beeston mart site and a busy dairy cattle business was developed with the partners of the firm anxious to compete with the big centres of Crewe and Chester, each equidistant to the east and west. Beeston was the first mart in the county to hold all TB Attested Sales and also the first to introduce the weighing of store cattle and, later, cull cows. In 1960, they were advertising that, every Monday, they had 1,000 fat cattle, sheep and pigs, 750 breeding and store pigs and 2,500 poultry and, every Friday, 120 dairy cattle, 250 cull cows, 250 store cattle and 250 calves.

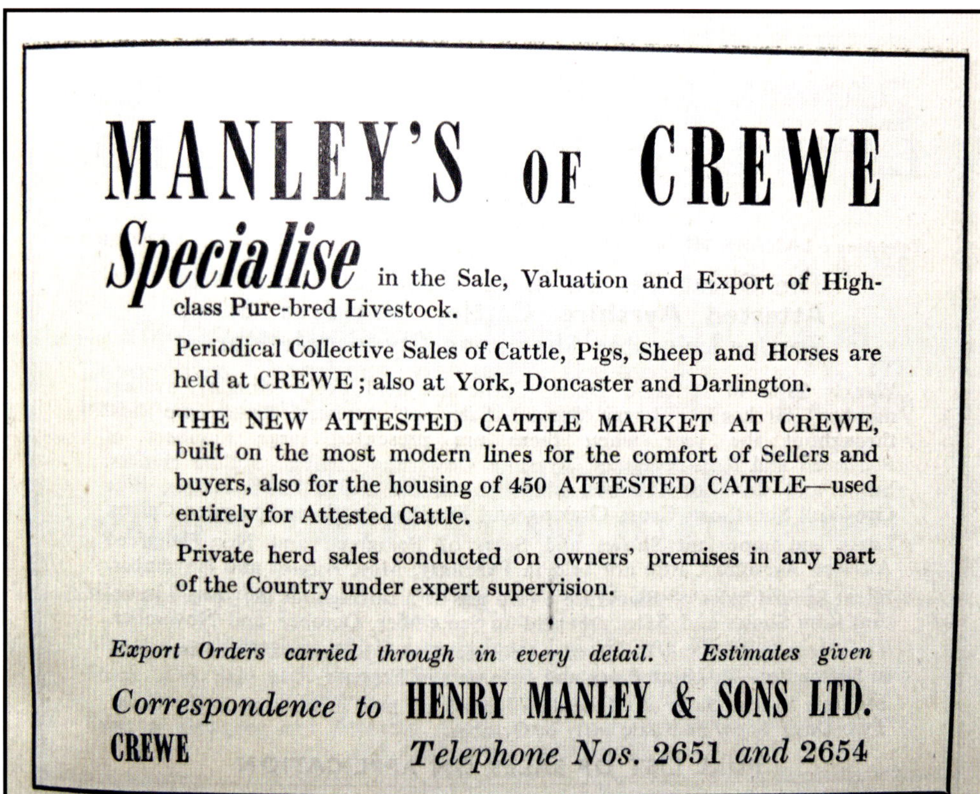

In **Crewe**, sales had commenced in 1872, two years later than at Beeston, when a William Hill conducted them in a field behind the

Royal Hotel. The sales were firstly held monthly but the proximity of the railway station ensured their success and they quickly became weekly and attracted local investment and interest, not least from the directors of the London and North Western Railway Company who participated, in 1882, in the establishment of the Crewe Cattle Market and Abattoir Company. New premises, at Gresty Road, with special rail access were erected and opened in 1883. The auctioneers appointed by the new company were J.C. & H.H. Etches but very soon the owners found it preferable just to let the market to the auctioneers and let them get on with it; Messrs Etches took the tenancy along with Henry Manley.

HENRY MANLEY & SONS, LTD.

(Established 1861)
AUCTIONEERS, VALUERS AND SHIPPERS OF PURE-BRED LIVE STOCK.
CREWE AND BRANCHES
ANNOUNCE SALES AS FOLLOWS:—

TO-MORROW, WEDNESDAY, JULY 24th (2 p.m.)
AT CROOKS FARM, WIGAN LANE, COPPULL, NR. CHORLEY, LANCS.

30 Attested British Friesians

being the Entire Bridgemere Herd for Mr. J. B. Dutton. Also 1,368 R.I.R. cross pullets (hatched Feb. to April) and Farm Equipment at 11.30 a.m.

TO-MORROW, WEDNESDAY, JULY 24th (12 noon)
IN THE ATTESTED SALEYARD, CREWE

16 Attested Pedigree Jerseys

being an attractive entry of Cows and Heifers, either in-milk or forward in-calf.

AND

69 Attested Guernseys

consisting of a very fine selection of 58 Pedigree and Non-Ped. Cows and Heifers, either recently calved or well forward in-calf; and 11 Maiden Heifers. Included is the usual selective consignment from Messrs. R. & J. C. Eddy.

FRIDAY NEXT, JULY 26th
(Judging 9.30 a.m. Sale at 11 a.m.)
IN THE ATTESTED SALEYARD, CREWE
The North-Western British Friesian Breeders' Club's Show and Sale of

125 Attested British Friesian~

comprising 84 Cows and Heifers, either calved or close to calving; 26 Maiden Heifers and 15 Highly Bred Young Bulls. There will be a good percentage of Young Cows and Heifers, at or close to profit, and many well-known herds are represented.

MONDAY NEXT, JULY 29th (12.30 p.m.)
IN THE ATTESTED SALEYARD, CREWE
Weekly Sale of about

160 Attested Commercial Dairy and Store Cattle

being Friesians, Dairy Shorthorns and Various Cross Breeds and comprising Cows and Heifers either calved or near calving; Stores; Calves.

AUGUST

2.—Attested Ayrshires at Crewe.
5.—Attested Commercial Dairy and Store Cattle at Crewe.
7.—Attested Jerseys at Crewe.
7.—Attested Guernseys at Crewe.
7.—The Yorkshire Pedigree Pig Breeders' Association's Show and Sale of Pedigree Breeding Pigs at York.
12.—Attested Commercial Dairy and Store Cattle at Crewe.
15.—The Crewe (Cheshire) Association of Shorthorn Breeders' Show and Sale of Attested Dairy Shorthorns at Crewe (with Messrs. Thornborrow & Co.).
16.—Attested Friesians (Registered and Non-ped.) at Crewe.
19.—Attested Commercial Dairy and Store Cattle at Crewe.
21.—Special Periodical Draft Sale of Attested Guernseys for Messrs. R. and J. C. Eddy at Crewe.
23.—Attested Ayrshires at Crewe.
23.—Pedigree Large Whites and other Pure Breeds of Pigs at Crewe.
26.—Attested Commercial Dairy and Store Cattle at Crewe.
28.—Attested Jerseys at Crewe.
28.—Attested Guernseys at Crewe.
30.—The North Western British Friesian Breeders' Club's Show and Sale of Attested Friesians at Crewe.
30.—Special Show and Sale of Pedigree Landrace Pigs at Crewe.

SEPTEMBER

2.—Attested Commercial Dairy and Store Cattle at Crewe.
6.—Attested Ayrshires at Crewe.
9.—Attested Commercial Dairy and Store Cattle at Crewe.
11.—Attested Jerseys at Crewe.
11.—Attested Guernseys at Crewe.
13.—Mr. T. C. Bell's 4th Annual Autumn Sale of 250 Attested Dehorned Scottish Ayrshires (removed from Scotland) at Crewe.
13.—Special Show and Sale of Pedigree Essex and Wessex Saddleback Pigs at Crewe.
16.—Attested Commercial Dairy and Store Cattle at Crewe.
18.—The Yorkshire Pedigree Pig Breeders Association's Show and Sale of Pedigree Breeding Pigs at York.
20.—Attested Ayrshires at Crewe.
20.—Special Show and Sale of Pedigree Large White Pigs at Crewe. (Special fixture for Progeny Tested Stock).

SALE NOT PREVIOUSLY ANNOUNCED

Oct. 30th.—Selection of T.T. Pure-Bred Landrace Pigs from the "Eaton" herd, property of The Executors of the Duke of Westminster, G.C.V.O., D.S.O., at Eaton Home Farms, Aldford, Nr. Chester (same date—Messrs. Thornborrow's conduct sale of selection of Dairy Shorthorns).

CREWE (main railway line—London to Scotland) maintains its position as the most Modern and Largest Market for the sale of Attested Cattle in England and Wales. A market entirely segregated in the midst of the largest Dairying Area. Average of 3 sales held weekly. Private herd sales accommodated in their entirety.

CATALOGUES, ENTRY FORMS, ETC., FROM
HENRY MANLEY & SONS, LTD., CREWE (Tel. 4301-2-3).

Manley, the son of a wheelwright and born in 1828, first held his auctions at Aston, by Wrenbury, in 1861 in premises adjoining the Salamanca Hotel which were reckoned to be the first purpose-built auction mart in Cheshire; he was also selling at Whitchurch, as were Etches, when he became involved with Crewe.

Etches and Manley traded together for some years but were then taken by surprise when, in 1895, the Crewe Cattle Mart and Abattoir Company terminated their arrangement and leased the mart to Frank Lloyd of Wrexham. Henry Manley then had to hurry to build his own new mart which opened, on the other side of Gresty Road in February 1896; this later became known as the Middle Market. Manley took his four sons into partnership in 1900 and died in 1903[38]. Probably in the late 1890s, a third Crewe Mart was erected by J.M. Barker and Company, latterly Mr Massey Barker, who also sold at Northwich. Three firms and three marts were then in competition in the town.

By 1930, Henry Manley and Sons had brought all three marts under their control and the company chairman was Ernest Wright[39] who had previously been with their opposition, Frank Lloyd and Sons. Robert and Hugh Manley were both directors and they had established what was probably the premier firm of livestock auctioneers in England with four rings in Crewe and branches at Whitchurch, Nantwich and Market Drayton together with a pedigree sales office in Southampton Row, London. Dairy cattle were a mainstay and their sales at the Horse Repository at Crewe, opened in 1901, were renowned with maybe 1,300 foals over two days at the Autumn event, regular Hackney and Shire sales and sales of Greyhounds on a Saturday. Immediately post World War II, Manley`s had the foresight to make a major investment; rebuilding the Barker Market with a purpose-built pedigree cattle centre and then undertaking a major re-design of the Middle Market. Some £100,000 was spent, a very substantial sum at the time.

Manleys had strong competition, for pedigree sheep sales in particular, from Alfred Mansell of Shrewsbury. Mansell was secretary of the Shropshire Sheep Breeders Association and this involved him, also, in the world-wide export business. Mansells were progressive people and, as early as 1900 Shrewsbury Mart store cattle sales were advertised as being "on the live weight system".

---

38 Henry Manley left an estate of over £10,000 – an unusually substantial figure for an auctioneer. His executors were his sons, Robert and Edmund.

39 The name of Wright crops up persistently in Cheshire auctioneering annals. Joseph Wright of Beeston, Ernest Wright of Crewe, Leonard Wright of Chester and Stanley Wright who was principal at Warrington Mart. Strangely, it is believed that none of these men were actually related.

For some four centuries there was a cattle market in the centre of **Chester** but the first auction mart appears to have been the Ermine Cattle Market on Hoole Road which was situated, as was essential, near to a railway station. An article in the Cheshire Sheaf in 1921 records that "on the left-hand side of Flookersbrook stands the Ermine Hotel, an old-established hostelry, with its extensive stabling and auction mart"; at the time that that was written the mart had been there for some fifty years, having been established by 1870. One wonders if the mart accommodated the visit, in 1871, of Day`s Crystal Palace Menagerie whose exhibits at the Ermine Hotel included "The Umbrella Elephant from Abyssinia". The first auctioneers at the Ermine Mart were Pickering, Sons and Woodfin, later, by 1882, Pickering and Nightingale and then Mr Pickering operating alone. Pickerings also sold in Greasby, Hooton, Moreton and Flint.

Beeston Castle Auction Mart

By 1878, in Chester, R. H. Glover was advertising fortnightly sales of cattle, sheep and pigs at Bowling Green Cattle Mart. This mart may have been relatively short-lived but Glover also held horse sales in the Grosvenor Hotel Paddocks. A small sale of dairy cattle was advertised in this same year at The Mart, Foregate Street, Chester by Churton, Elphick & Co. They had a fairly extensive business and conducted many on-farm sales but their "Mart" perhaps tended to be used more for sales of chattels and general effects than livestock.

In 1884, Chester City Council arranged the building of a publicly owned mart between George Street and Gorse Stacks. This was in two sections; the Smithfield which was let to John J. Cunnah, later Cunnah and Roberts and the Agricultural Hall which was taken by Sidney R. Fearnall. Thus, as in Crewe, there ended up being three markets in the town competing for business. Later the Ermine Mart would close. This left the two council tenants in opposition to each other until, sometime in the 1920s or 1930s, Roberts died, Cunnah retired and Fearnall gave up his tenancy of the Agricultural Hall. Leonard Wright, who had worked at the Smithfield since 1908, then tendered for the lease of both sections of the market. He was joined by Leigh Sadler who had worked with Sidney Fearnall and the new firm traded as Leonard Wright and Co. Wright also held weekly sales at Greasby and Hooton, both of which had previously been branches of the Ermine Mart. He was also secretary to the Frodsham Farmers Auction Company. Sales at Frodsham were on a Wednesday at 2pm. The mart was unique in that it had a square ring. Les Foster sold there for many years.

Chester Market went on to develop successfully over the next thirty years or so and by 1960 the weekly advertisements were for Tuesday, with sales of 150 prime cattle, 600 fat sheep, 1,200 pigs and 500 calves, and for Thursday with 400 fat cows, 300 store cattle, 100 dairy cattle and 2,000 store sheep. However, by later in the 1960s, the business was suffering from an ageing partnership, a town centre site and the loss of agricultural land on the Wirral. Robert Brown and John Minshall had become the principals and they saw the advantages of a merger with their Beeston rivals and so Joseph Wright and Leonard Wright and Co. amalgamated. With the demise of the marts at Frodsham and Hooton, the new firm attempted to prolong business in Chester by taking on the new market at Sealand which was built by Chester City Council and opened in 1971; but the rent was high as were the running costs and staunch Chester supporters from the south and east of the city tended to trickle away to Beeston which was now of course part of the same firm. Nonetheless the mart continued to do fairly well for some years with a strong following from North Wales. Wright Manley relinquished their lease on Chester Mart in 1995 and Frank Marshall and Co. stepped in; things went well for about a year until the BSE crisis and then, in turn, the Foot and Mouth outbreak; a short reopening did not result in adequate support and the mart closed in 2003.

The overall success of the Wright business at Beeston Castle and Chester put some pressure on Crewe and it was not long until discussions between Wright`s and Manley`s took place. The firms merged to become Wright Manley in 1979[40]. In time, both Chester and Crewe would close with the business concentrated on Beeston and, in 2014, a further significant merger took place when Wright Manley joined with Frank Marshall of Chelford to become Wright Marshall.

Crewe the rebuilt Barker Mart, c1951

Chelford was relatively late in getting the benefit of an auction mart. A Mr Thomas Wilson had held periodic sales in the town but it was not until 1910 that the Chelford Cattle Market Company was launched with only £450 of capital. Land was rented from Sir George Dixon and the mart opened in 1911. These modest premises had a sale ring, ties for forty-two cattle, two boxes for calves, three pens for loose cattle and sixteen pens for pigs and sheep. "There was an office heated by oil stoves and lit by oil lamps, where money changed hands and three pens for stock loading plus a long platform, which was used to unload horses and to handle milk churns. A timber building on the site was used to stable horses overnight after they brought the mail from Knutsford. Water came from two storage tanks owned by the railway company fed by a wind pump". The Cattle Market

40 As far back as 1960 the two firms had been co-operating as "joint auctioneers" in Whitchurch Mart.

Company bought the freehold of the site in 1926.

John E. Braggins was the auctioneer and, in 1933, he bought the assets of the Chelford Cattle Market Company and, some two years later, took Frank Marshall as a partner. Marshall had worked for Braggins since 1917 when he was thirteen. Braggins[41] retired in due course and Frank R. Marshall and Company was established in 1947. The site was extended in 1952 and the mart was progressively improved to incorporate pens and buildings brought from Knutsford and Altrincham Marts when they closed. Over the years, Marshall`s acquired the businesses of Marsh & Son of Northwich, J. Mainwaring and Co. of Warrington and J. K. Hill Hampson of Buxton. A very successful centre was established at Chelford, to include important sales of pedigree sheep and cattle together with sales of poultry and horticultural produce, but, shortly after the merger with Wright Manley, the mart lease expired, the owners of the site insisted on re-development and Chelford Mart shut its doors in 2015.

John Hockenhull and his son, Sydney, were among the largest sheep graziers in Cheshire, often having as many as 10,000 head out on farms. Many of these were destined to be sold in Chelford.

Whitchurch Smithfield, 1893

The mart at **Malpas** opened in 1872 at Hampton, close to the railway station. Booth Hewitt was the auctioneer. By 1890 Mr Frank Lloyd was holding substantial weekly sales there and, in September of that year, 60 cattle, 300 sheep and 300 pigs were penned. By 1920 the sales were held jointly with Jones and Son of Wrexham who later acquired Lloyd`s business when they had marts in Wrexham, Llangollen, St Asaph and Oswestry. Malpas was operated as Jones, Son and Staff, later

41 John Braggins & Co., the name continues as Meller Braggins, surveyors & estate agents.

under the management of Douglas Brereton, a popular auctioneer who, in 1980, was averaging about 130 prime cattle, 500 sheep and 100 pigs every Saturday; the site was sold for development in 1990. The company continues as Jones, Peckover and Co in Wrexham and at St Asaph where John Brereton is the managing partner of the mart business.

Lloyds, auctioneers, opened a mart in the centre of **Congleton** in the late 1880s. They traded there until the late 1920`s when their lease ended and, for a very short period Congleton Mart was operated by Oakes Ash who ran marts at Leek and Newcastle in Staffordshire. In 1931, Billy Whittaker, who had worked for Lloyds since the early 1920`s, took the lease on the mart and set up in business with an office in Kinsey Street.

In the early 1950s, Geoffrey Biggs, who was a land agent and who ran the Antrobus Estate in Cheshire merged his practice with that of Billy Whittaker and the firm of Whittaker and Biggs was formed. Geoffrey's son, Mr Roger Biggs entered the practice and in 1964, Billy Whittaker retired. In 1964 John R. Knott, who was an auctioneer at Chelford with Marshalls, moved to Whittaker and Biggs and in the same year the mart moved from the town centre to a new market site on the northern fringe of Congleton on the Macclesfield Road. During the 1960s and 70s and into the early 1980s Congleton was a successful one-day market selling up to 300 to 400 barren cows each week, 150 fat cattle, 500 fat lambs, 100 pigs and circa 50 calves with periodical sales of store cattle and machinery on the site. As with other Cheshire marts, the decline of dairy cattle sales as a result of health accreditation and milk quotas hit Congleton hard, as did the BSE crisis but, as John Wilson of Whittaker and Biggs points out, "the killer blow was Foot and Mouth in 2001". Numbers had been improving until then but there was little support when the mart re-opened in 2002 and it closed in 2003.

Chester Mart with David Fisher on the rostrum

Leonard Wright opening the new Attested Market at Gorse Stacks, 1955

It is astonishing that only some twenty weekly marts still operate in the three counties of Yorkshire, Lancashire and Cheshire and, from over twenty-five marts in the latter county, only Beeston Castle remains. Now led by Gwyn Williams, past president of the Livestock Auctioneer's Association, and Nigel Eckersley with a most successful team of Clive Norbury, Johnathan Farrall, Simon Lamb and Jonty Cliffe, Beeston provides a wonderful service for the sale of pedigree and commercial livestock.

Plans are underway to replace Beeston Castle, and the recently closed Chelford Mart, with new premises but the planning process is proving tortuous. It is a tremendous change from the days when so many towns or rural railway stations were thought to merit an auction mart.

By the late 19th century, there were marts or "Smithfields" as they were often referred to, of one sort or another, as follows. Some would be substantial and optimistically erected premises; others, doubtless, would be but a cluster of pens.

Altrincham	Possibly John Braggins, then Frank Marshall. On closure, buildings & pens relocated for use at Chelford.
Beeston Castle	See above
Boughton Hall Station	J. Brown
Calveley	Booth Hewitt
Chester	See above
Chelford	See above
Congleton	See above
Crewe	See above
Frodsham	Frodsham Farmers Auction Co., Leonard Wright, Secretary (behind the Bear`s Paw pub)
Greasby	Pickering (of the Ermine Market). Sidney R. Fearnall. Wirral Auction Co. Leonard Wright & Co. (at the New Inn)
Handforth	Seasonal mart with sales in Spring and Autumn
Hooton	Pickering (of the Ermine Market). Cunnah & Roberts (at The Old Smithfield). J.P. Carter. Wirral Auction Co. (at The Smithfield) Leonard Wright & Co.
Knutsford	A Farmer Co-operative. Operated by John E. Braggins (see also Chelford). On closure, buildings and pens relocated for use at Chelford.
Macclesfield	Thomas Turner, 1875. Turner & Son, 1899 to 1970s
Malpas	See above
Marple	At Brabyn`s Brow, 1909 – 1968. Brady`s of Stockport
Nantwich	Henry Manley & Sons. T.W. Fowles at Nantwich Swine Market
Northwich	Marsh & Son. J.M. Barker & Co. Henry Manley & Sons
Moreton	Pickering & Nightingale
Pulford	Pickering

Stockport	Cattle market established 1850s, later moved to Portwood. "Large" sales reported in 1900. Probably closed circa World War I
Tattenhall Station	Booth Hewitt. T. Broster. F. Lloyd
Warrington	See Chapter 14
Whitchurch, Shropshire	Henry Manley & Sons. Frank Lloyd & Sons. Joseph Wright. (Closed 1975)
Worleston Station	Booth Hewitt
Wrenbury	1860, Henry Manley & Sons. Wrenbury Farmers` Auction Co. Joseph Wright. (Largest throughput in Cheshire as a World War II collection centre)

At Chelford Dairy Show, 1955

Numbers were never immense at the various marts and a brief look at September sale results for some of the marts over the period of the 1890s through to 1910 indicates the following approximate weekly numbers of fat cattle sold.

Beeston Castle	Booth Hewitt	80
Chester	Cunnah & Roberts	75
Chester	Churton Elphick	60
Chester	Pickerings	50
Chester	J.E. Davies	50
Crewe	Henry Manley	65
Chelford	Thomas Wilson	30
Frodsham	Farmer's Mart	50
Hooton	Wirral Auction Co	90 (incl stores)
Greasby	Wirral Auction Co	60
Moreton	Pickerings	40
Congleton	Lloyds	50
Tattenhall Station	Lloyds & Brosters	70

Congleton Mart, 1997, John Wilson Auctioneer, Mr Dale owner of champion, Mr Knott senior partner, Mrs Richardson presenting prize

Sheep Sale at Congleton

At the author's Scottish marts at Annan and Dumfries, there were many buyers of dairy cattle from Lancashire, Yorkshire and Cheshire, including a few very loyal friends from the Cheshire area; in particular, Philip Buxton and his friends from Macclesfield, the Moss family from Congleton and Syd Biddulph, also from that airt. There was also Walter Holgate from Preston, who first came in 1945 and for several years never missed a sale, and Benny Mills from Otley, to mention only one or two of many. Mills and Buxton were the first men to buy dairy cattle at over £100 per head. Reg Roberts from Tarvin at Chester bought huge numbers and was always a good friend and a pleasure to meet, along with many other dealers such as John Bailey and Charley Bailey, who incidentally were not related. All were men of great integrity, but all liked a good luck penny from the seller.

CHAPTER XVIII
1955 TO 1990

The post-war return to a free market had been combined with a grading system to allow for a government guaranteed price to be achieved by a deficiency payment scheme covering cattle from 7cwts to 15cwts liveweight and lambs from 17lbs to 70lbs estimated deadweight. This worked very well with only stock sold by auction being used in the government calculations, although stock sent direct to approved deadweight grading centres did also receive the deficiency payment. Around seventy per cent of fat or prime cattle and sheep were, at that time, sold by auction.

In 1954/55 the average guaranteed price was £7.55 per cwt for cattle and 16p per pound for sheep. The scheme provided that when average prices were below the guarantee the Exchequer made good the difference, but when they were above it the farmer retained the benefit. It still, therefore, paid to get the best possible price at your mart. Under the scheme, the auctioneer was responsible for ensuring that every graded animal, once sold, had a hole punched in its ear.

A small calf subsidy and a small ewe subsidy were also paid to producers at that time which provided for a reasonably stable industry in which auctioneering could flourish.

Prior to the war, in 1939 there had been around a hundred auction marts in Lancashire and Yorkshire. Many were owned by the local authorities and were often operated by more than one auctioneering firm; indeed, there were over one hundred and fifty separate firms. Hull Market led the way with twenty four firms of which only eight re-commenced as livestock auctioneers when business re-started fifteen years later, although some of these were amalgamations of two pre-war competitors. York had supported six firms and the advertisement on page 55 shows how they operated.

Many towns had two, three or four auctioneering firms, but by 1960 some of these had fallen by the wayside with the remainder carrying on with varying success. The generally healthy post-war state of the auctioneering business continued for about thirty years but with many changes. The greatest of these changes were decimalisation and metrication. In 1968 auctioneers finally accepted that it was inevitable that they must change their age-old patter with the introduction of decimal coinage. In consultation with the Meat and Livestock Commission they prepared for D Day on 15th February 1971. Agreement had to be unanimous amongst sellers and buyers when shillings disappeared and new pence took their place. The well-worn Donkin's ready reckoner books prepared by a Northumberland auctioneer a century earlier were consigned to the bonfire. It all went very smoothly and encouraged auctioneers for the next big change which was from hundredweights and pounds to kilos. This change was more costly as all weighbridges had to be re-calibrated and many older ones had to be replaced at substantial cost.

Under the old regime, the County Weights and Measures Inspector would come twice a year to check the machines. A load of fifty-six pound weights had to be provided by the auctioneers as a test unless you lived in one part of the west of Scotland where the butcher's delivery pony was weighed and then, with a certificate attached, used as a test weight for the local mart.

We also had the change from acres to hectares, which farmers have never been able to accept as it immediately appeared to reduce the size of their fields and farms. 220 acres sounded so much more impressive than 88 hectares.

The changing pattern of business was demonstrated by the decision, in 1962, of the British Transport Commission to withdraw livestock facilities from all the small stations which had provided such a handy (obliging – convenient – speedy) service for farmers since the railways began. This, in time, led to the total demise of livestock transport by rail.

As mentioned in the previous chapter, a major problem for auctioneers was bad debt and this remains a problem to this day. This is not a new issue and, back in 1929, the Auctioneers Institute had sent the following letter to livestock auctioneers and an early reply was requested to the following questions.

a) Whether you approve of the issue of a list of dealers and butchers with whom members should deal on cash terms only?

b) Whether you will undertake to regularly furnish to the Institute the names of all butchers and dealers who fail to make due payment for their purchases?

Needless to say such action was not agreed as each firm had a few friends who got special terms and, in any case, it has always been difficult to get all auctioneering firms to agree on anything decisive.

On re-opening the fat stock sales in 1954 most buyers took a week's credit and a few special cases got fourteen days. But then came the supermarkets who took from four to eight weeks to pay the abattoir and were such large customers that the wholesalers depended on them and lost control. Then the marts allowed three or four weeks credit and they, in turn, lost control. To compound the problem, sections of the industry, the ewe trade in particular, became dependent on certain buyers who proved to lack financial staying power and who caused some very big losses to auction marts. The people who did not pay effectively stole large sums of money from the mart shareholders who, as always, protected the sellers from the loss. It is to be hoped that the new generation of professional auctioneers will have been taught that when finding a new buyer for their sales they will also ensure that they can pay and that, on the first sign of default, a blind eye is not turned - delay is a mistake.

Auctioneer, Vincent Sidall booking in pigs at Wadsley Bridge Mart, 1962

In 1968 parliament had set up the Meat and Livestock Commission to improve all procedures between the farm and the kitchen table. There were three statutory committees, Production, Distribution and Consumption. Five years later, among the auctioneers on the twenty-four man Distribution committee were Dick Harrison of

Carlisle, H. Lacy Scott of Bury St Edmunds and the author of this book from Annan; together with Yorkshire and Lancashire men, Reg Stephenson, Eric Dee and Richard Turner; a splendid group of friendly people from whom I learned a great deal. The MLC staff who advised on the design of the marts and slaughterhouses, not always with the most practical solutions, spent a lot of money on livestock improvement schemes. On the technical side they did some good work in eradicating the warble fly in England which had spoiled so many good valuable hides and Joe Read, head of the sheep development service, laboured to improve our multiplicity of sheep breeds. Lacy Scott, a shrewd auctioneer, was a good fighter in committee for the marts which still handled around two thirds of the cattle and sheep which were sold. His mart has now closed.

Marketing was an important part of the Meat and Livestock Commission remit and was subsequently handled by EBLEX, now AHDB, Beef and Lamb[42].

I felt that the MLC staff were inclined to be in favour of dead weight marketing and a careful watch on them was essential. The Commission was financed by a levy on all stock slaughtered which is still collected today and much of which is spent on works to improve, promote and defend the prime stock industry. The levies today are £5.40 for cattle with sheep at 80p. Of this, three quarters is paid by the farmer with the auctioneer acting as the collecting agent.

AHDB undertake research, provide market information, encourage exports and publicise the benefits of meat throughout England and Wales whilst also doing a good job in promoting exports which also helps our national balance of payments. Quality Meat Scotland does the same job in Scotland with additional responsibilities for pork, but complain that, as many Scottish animals are killed in England, their English counterpart gains revenue unfairly. Needless to say I am sure the board of QMS would be improved by the appointment of an auctioneer.

There is a constant tendency for all these self-perpetuating semi-official bodies to try and avoid supporting the auction marts; just because something is not new, they think, somehow, that it is not good. In this and other areas, they are not necessarily acting in the best interests of the farmer whilst cheerfully spending his levy money.

The Livestock Auctioneers Market Committee for England and Wales

42 EBLEX: English Beef & Lamb Executive. AHDB: Agricultural & Horticultural Development Board.

represented the auctioneering companies and fought their corner. In about 1970 they obtained a 15% increase in the certifying fees paid to the auctioneers by the Ministry of Agriculture, just as the Ministry were handing over some of their responsibilities to the Meat and Livestock Commission. Although the latter body enjoyed good relations with the auctioneers, they favoured deadweight, rather than liveweight certification. Bulls and calves were an immediate example of this as grading of these moved to the abattoir. The MLC 1970 report noted a general trend in that certification in auction marts had, since 1955, dropped 14% for fat cattle to 54% of the total and dropped by 3% for fat sheep to 60% of the total. The costs of the Market Committee were assisted by each company contributing at least £50 annually for each auctioneer selling or grading in the market. A few firms tried to avoid paying but they were generally pursued by their local county committee.

During this period of peace and prosperity the attendance of local retail butchers and several smaller wholesalers at every mart sale of fat stock ensured competition and fair prices. I was reminded of this at a recent lunch with a Lancashire farmer up to buy store lambs at Longtown. His family had been butchers and now in their town there was only one butcher's shop instead of seven after the war. This, in turn, brought back memories of Mr Aitken from a mill town who would send a lorry to block off the main street as four cattle and ten sheep from the mart were persuaded to cross the pavement and squeeze down a very narrow "wynd" or "close" to a small shed behind the shop where execution took place. Such practices were soon to be prohibited by bureaucracy and the closure of many small slaughter houses contributed to the decline in prime cattle sales. In 1962 there were 415 slaughter houses in Yorkshire and Lancashire but now only a few remain.

Traditionally those retail and catering butchers who did not have their own abattoirs were served by a large number of wholesalers who had their own slaughter house or used the public facilities provided by the local authority.

Each town had at least one established wholesaler including Bert Noblett at Preston, Goddard of Huddersfield, Ashworth of Bacup, Slinger of Accrington, Pooley of Chorley, Newton of Southport and Garlick of Lancaster, Birtwhistle, Butler, Jones and Wharff from Manchester, Foster of St. Helens, Booth of Bolton, Winder of Blackpool, Blackledge of Great Harwood, Wildblood & Swires of

Leeds, Brian Smith of Rochdale, John Simpson, Wooleys, Woodheads and many others.

In the cities, there were great wholesale meat markets with a dozen or more stalls for the fresh meat men and further stances for the large international meat importers to sell their wares.

George Fieldhouse

Among the most enterprising of the many northern meat wholesalers was George Fieldhouse, a very hard-working man who travelled the marts in a large and flamboyant American automobile. His firm, Fieldhouse and Coop built, in 1960, a state of the art modern plant at Dukinfield. They were loyal supporters of the auction mart system, from Norfolk to Scotland, attending regularly to select the animals with the carcasses to suit their wide range of customers; and they were good payers too.

George argued that the supermarkets would be ruin to us all and particularly to the medium-sized wholesalers and the local butchers. As always, he took his own advice and sold out to Louis Edwards in 1975; a wise and far sighted decision.

In the 1970`s killings in the city abattoirs of Liverpool and Manchester were in steep decline as these premises were superseded by modern slaughter houses in out-of-town locations with cutting rooms supplying the supermarket chains direct rather than selling through the meat-market wholesalers. By the mid 1980`s slaughtering in these two great cities had ceased.

In this period, prices of prime or fatstock moved steadily upwards governed by an annual price review involving the government and the farmers unions. By January 1972, the guaranteed price had reached £12.64 per live hundredweight for cattle and twenty three pence per pound estimated dead carcase weight for lambs. To qualify for this guarantee, Irish cattle had to have been in the UK for at least ten weeks. The animals were weighed and graded[43] and the auctioneer certifying officer had to ensure that the task was carried out properly and that each animal was permanently marked at the fall of the hammer, so that the guarantee couldn`t be claimed more than once. One or two marts were reputed to underweigh their lambs in order to be able to quote a higher price per pound in order to attract sellers.

A massive reduction in the number of slaughter facilities at this time had generally started in Brussels where the EU wanted to make life difficult for British exports, which continued as French farmers dumped lorry loads of lamb in the docks, and continued with demands that slaughter houses in Britain should be modernised at great expense. This was then extended by British civil servants to all the smaller killing houses, many of which could not comply and closed. The promises of larger units to kill for them proved unreliable and forced many butchers to buy deadweight, thus cutting out the auction mart weekly prime sale. By the mid 1980`s the large supermarkets were full of cheap imported meat to woo the housewife. The pressure to close local slaughter facilities was misguided and, in these days when it is fashionable to have short supply chains and locally produced produce, these facilities are greatly missed.

Auctioneers did receive a welcome European boost in 1980 when EEC grants to modernise plants processing agricultural produce were extended to include auction marts. This resulted in some being modernised and some others being replaced by new marts. One of the first new marts in England and Wales was Lampeter, opened in 1984, at a cost of £300,000 to which the EEC made a grant of £81,000. Sadly this mart is closed, like so many others.

43 A little rhyme used to run "This is a plea to the ministry graders; Please be fair to all meat traders".

In 1988 spiralling rent demands by local authorities forced the closure of markets at Leicester, Mansfield and a new post-war mart at Slough, with others fighting for survival.

CHAPTER XIX
CATTLE DISEASES

1991 was a welcome date for auctioneers when 47 years of guaranteed prices and grading of fat stock came to an end and the price in the ring was what the farmer got. The marts continued to weigh the stock and announce the weight but the buyer had to assess the quality, there was no more blaming the grader!

So that all was going well for most marts until 1996 when a suggestion that a human disease called CJD was caused by BSE in cattle. The bottom fell out of the beef trade, then the sheep trade collapsed.

BSE in cattle was, arguably with very little credible evidence, accused of having the potential to kill large numbers of people. This meant that carcases of over 30 months old cattle could no longer be sold and cull cows lost half their value causing great loss to dairy farmers and to the auctioneers who were forbidden to sell them, thus losing a substantial weekly income, which left them struggling to survive with no compensation. The media called this, "Mad Cow Disease".

It was suggested that sheep also might transmit the disease and immediately government restrictions were placed on sheep over a year old, the age to be ascertained by opening the mouth of every sheep and counting their teeth. Ewe prices fell by about 50% although no proof of disease transfer was ever found and it has taken some twenty-five years to get government to even discuss a relaxation of the rules for sheep. The cattle rules were relaxed earlier to help our native breeds, which mature more slowly and cast cows were again sold in large numbers through the marts.

By 2000, after a very difficult decade, auctioneers were looking forward to the new millennium with confidence, but this was to be short lived as, in February 2001, Foot and Mouth Disease reared its ugly head again. It was first recognised on a Northumberland pig farm, then stock moving from Hexham and Longtown marts spread it quickly throughout England, even reaching Cornwall in March.

It was an appalling time for all concerned with over six million animals slaughtered in the UK to ensure the eradication of the disease, including the losses in the following counties:

	Cattle	Sheep	Pigs
Yorkshire	54,739	300,389	19,049
Lancashire	21,233	77,883	1,544
Cheshire	2,584	6,714	916

Numbers of cases peaked in April and slaughter did not conclude until November. Only the Cumberland area really lost control of the disease with some 900 cases and the army having to be brought in to deal with a huge backlog of unburied carcases.

All marts would be closed for a whole year with no income, with the inevitable result that some did not re-open as their operators lost heart or saw this as a useful opportunity to get out of a business which had, perhaps, not been very remunerative for some years. Equally some marts were so located that compliance with more stringent regulation was impossible in practical terms and, for others, the temptation of a large price for re-development was just too great to ignore.

Auctioneers in diseased areas carried out large numbers of valuations, working long hours in the most trying situations of grief and mayhem. A year later, when trying to re-stock, farmers often found their compensation money to be a little short of what they needed and certainly not the gold mine that some critics of auctioneers had suggested, although there were certainly some exceptions. Unexpectedly, the Government challenged the scale of valuation fees which had already been agreed by stating that they were "per day" and not "per valuation". Two firms took the matter to court, Mitchells of Cockermouth and Wallets of Castle Douglas; both winning their cases and getting fair treatment for all the valuers but only after a long delay.

Unbelievably there was soon another lesser outbreak of Foot and Mouth triggered by a release from, of all places, the Veterinary Research Centre at Pirbright. This was soon controlled but, again with considerable disruption of the marts and marketing systems.

It must sincerely be hoped that it will be a long time before there is a recurrence of the disease and that it can be dealt with in a manner which is more humane to both animals and their owners and herds-people. Perhaps lessons have still not been learned as funding for disease outbreak controls have been cut at a time when new diseases such as Blue Tongue and Schmallenberg are appearing, the use of antibiotics is being curtailed and, above all, T.B. in cattle has grown to the terrifying numbers of 43,000 slaughtered in 2017.

A year on and perhaps some lessons have been learned as I read of two government initiatives. Firstly 5.7 million pounds to help farmers tackle BVD with more urgency and, secondly, a new test to detect Foot and Mouth Disease virus at an earlier stage before animals show the disease.

CHAPTER XX
THE ELECTRONIC AGE

During the twentieth century, auctioneers had loved and lived with their 19th century marts as they grew, sometimes haphazardly, to meet each changing demand. England abounded in those built before the age of cattle lorries but, suddenly, what had been taken for granted was no longer the way forward. Congested streets, grubby cafes, poor toilets and awkward alleys had to be replaced by palatial premises on new, often out of town, sites surrounded by car and lorry parks to rival the recent developments at Skipton, York and Clitheroe. Thus auctioneers depended upon their enemies, the supermarket chains, to buy the outdated sites of auction marts at high prices to finance the building of modern auction centres with large car and lorry parks, restaurants, toilet blocks, shops and professional offices; and with central heating to keep all their customers happy.

But there was to be little room for complacency as problems with disease, welfare, traceability, thieves and bureaucrats arose on all sides.

Unimaginable changes had taken place in livestock farming as the staple crops of hay and turnips almost disappeared, as did so many of our native breeds as Continental bulls and rams came to dominate our meat production. Away from the farm, motor lorries and trailers closed livestock transport on the railways and made drovers redundant. The trusted retail butcher, a mainstay of the weekly fatstock auctions, became a rarity in the face of the ever-expanding supermarkets who cared little for home produced meat but only for price.

Pigs had always been a part of mart sales, but the trade became very specialised with only a few major players. Pigs were now rarely sold at auction, an area of the business which had been immensely important in many Yorkshire marts with a thousand or so pigs at many weekly centres. A government order controlling the movement of younger calves made a big difference to some marts whilst control on movement of poultry resulted in yet another loss of income.

To offset these losses and with an eye to the future, some auctioneers turned their minds to technological solutions.

The lifestyle and culture of auction mart employees has been transformed by the electronic age. Whilst the marts used to employ mostly male clerks, with possibly a female typist, the majority of clerks and office staff are now female. Many of the young male clerks went on to become auctioneers and there was, certainly for a while, a shortage of young auctioneers. Traditionally, in most marts no one started selling until their mid-twenties having had a thorough grounding in all aspects of auction mart work but the trainee auctioneer now often starts selling in their teens with very little knowledge of the office functions. Neil McCleary, at Longtown, was not allowed to sell until the age of twenty four but went on to be a legend in his lifetime. Times have changed and encouragement of the younger auctioneers is crucial. At the 2018 world famous Stirling bull sales the senior auctioneers stood aside and the team in the rostrum were Ross Fotheringham age 27, Murray Rainnie age 30 and Raymond Kennedy age 27. Unthinkable in the days when the Fraser family held sway over MacDonald Fraser.

In the busy autumn season auction staff wives became "lamb sale season widows" as their spouses were at the mart from early, booking in the stock and dealing with enquiries in the office. The larger auction marts sometimes had a dedicated clerk calculating payments to the vendors and writing cheques but at smaller marts, once the sale was finished, the office team had their main meal, a convivial occasion full of banter, and then set-to to check the vendors statements, calculate commission and write the cheques, which a director signed before they were put into envelopes ready to post. In the author`s marts, a team of five did the work; two people calculating and completing the vendors statements, double checking each other's calculations, two writing cheques and adding up completed sales ledger pages, and the last member of the team putting up the envelopes. This often took until ten or eleven at night with, probably, an early start again the next day. This was before the days of calculators and mental arithmetic skills were honed and essential. Once VAT was first introduced, even more complex sums had to be done in the head, calculating 8.5% vat on the auctioneer's commission.

When calculators were finally available they were expensive and treated with suspicion by the old cashiers. One large auction company after a board meeting decided to risk purchasing two to try them!

As today, the clerk beside the auctioneer had to be quick and able. Cattle were weighed in hundredweights and quarters and sold in pounds, shillings and pence. In the fat (prime) cattle ring the clerk had to record this whilst the auctioneer sold two or even three cattle per minute. To find the total price the clerk had a "ready reckoner" book of tables giving the totals for all the various combinations of price and weight.

Computers in the auction mart office have superseded all this and whilst the atmosphere and bonhomie have changed, working in an auction mart office is still an exciting life with variety, a buzz, hard focused work but time for a laugh and a chat. The farmers and dealers who come in are generally pleasant with local gossip. They rely on the office staff to advise on the correct paperwork and to sort out any issues.

The initial computers in auction marts were relatively simple number crunchers. The sale was recorded in the rostrum on sale sheets in the time-honoured manner and then the information was typed into the computer which printed out the purchaser's invoices and the seller's statements and cheques. This saved a huge amount of post-sale time writing up ledgers, paying out the vendors and ensuring that the income and outgoings balance with each other. However if the computer broke down the mart did not grind to a standstill.

By 2018 the computer systems had become very complex and very expensive. If any component fails the auction sale has to stop.

There are very few suppliers of auction mart computer systems as they are a very specialist item and attract only limited demand. Newline Auction Systems from Devon is a major supplier to many auction marts. The system needs constant change to "talk to" the national databases, link to social media and to work with the latest operating software but each change produces issues elsewhere in the software thus making each amendment frustrating and challenging for the staff. Devices become outdated quickly requiring constant investment in hardware.

Technology has reduced labour costs by more than the costs of having the technology and many of the latest bureaucratic compliance functions could not be done without computers.

In addition to processing the numbers, the computer system records all the movements of stock with the holding numbers, type and quantity and updates the central databases of Trading Standards and other government bodies. Farm Assurance status and disease status are also checked particularly with reference to B.V.D. and T.B. testing. The system produces the statistics for the market reports, prints out the cheques or uploads a list of payments to the bank for direct payment to seller's accounts. It produces lists of people who owe money and is used for credit control and bank reconciliation.

In the modern auction mart, the entire system is integrated with weight input at the weighbridge, prices and buyers entered live in the rostrum and the production of reams of paperwork for buyers and sellers in the office. The paperless society is a long way away in the livestock world.

Auction marts now need back-up generators on site because a power cut is a disaster as the sale just stops. There is no manual system and few left with the skills to process a sale manually.

An auction mart computer system with a main server, which is a large computer that holds all the data and is then accessed by everyone from their laptop or terminal, and 5 or 6 terminals along with the software now costs in the region of £40,000 with an annual maintenance, updating and hardware renewal cost of £15,000 depending on the size of the mart.

Social media particularly Facebook, Twitter, Instagram and websites have become very important in the promotion of the auction market with traditional newspaper and magazine advertising playing a much reduced role.

A picture of a high priced animal can be viewed thousands of times in an hour. Videos of auctions and market reports ensure a surge of visitors to the site and can be uploaded to Youtube. Many marts put their ballot for the weekly sale on Facebook so that vendors can arrive at the right time to see their sheep sold.

The importance of social media in the 21st century is brought home to me every Thursday at Longtown when we hold our weekly sale of prime stock. The Thursday report on Facebook is viewed around 2,000 times every week by farmers and others seeking up to the minute information and guidance. Our website had 975,686 views in 12 months of whom 73.7% were returning visitors.

The auction mart's website is now of vital importance, providing up to date information on sales, stock available privately, photos, catalogues and market reports. Many newspaper adverts now refer clients to the website for further information. The pages of auction mart adverts in the farming press are getting much reduced with local paper advertising slashed. Many publications have disappeared including the free papers such as "Big Farm Weekly" which were so common in the 1990's.

A new website can cost between £2,000 and £10,000 depending on its complexity, with a small annual maintenance charge. Most are now written so that the auction mart staff can input the information without having to pay a specialist website operator.

Beeston Castle, Beltex Sale, 2018

"A five-gallon cow" was a good dairy cow and that was all that was traditionally announced by the auctioneer with all other selection being by eye. Gradually records of yield, butterfat and lactation period became commonplace in the dairy world. The essentials of the electronic age can include "estimated breeding values", scrapie resistance in sheep, Johnes risk level, T.B. testing interval and more recently genomic testing results with a D.N.A. map of the animal giving its "expected progeny differences" which covers a broad spectrum from hair colour to fat cover to docility, ease of calving, forecast growth and milk production and many other traits. Pedigree catalogues now provide detailed information on the sires and dams going back over a few generations with the latest genomic test results. This is complex information, where an inaccurate figure can paint a very different picture thus putting together and checking all the data is time consuming, but fascinating for the dedicated pedigree auctioneer and his team as they contact prospective buyers encouraging them to attend the sale or to log in online to study the information and to bid by telephone or remotely.

The complexity of the catalogue was demonstrated in Wright Marshall's "Beeston Blue" catalogue of March 2018 with some five pages of explanation of the "British Blue Group Breedplan EBV's". QR codes were also included for every lot permitting rapid direct access to the Wright Marshall website for photos and full pedigrees with the catalogue description limited to the Sire and Dam. In other catalogues, particularly for bulls, up to eight generations can be listed with brief statistics for each. QR codes are a symbol that can be scanned using a mobile phone that provide a link to a web page that then opens on the screen on the phone.

In the early 1990`s, with the online world expanding rapidly, some firms made an early and, in retrospect, ill-advised venture into electronic auctions for prime stock.

There were four electronic systems, the first being "Ease". This was an imported system purchased by Aberdeen & Northern Marts, who intended to franchise it to a limited number of companies to expand their scope of business out of the north of Scotland. Following the initial success of "Ease", a consortium led by Harrison & Hetherington Ltd., of Carlisle and Stephenson & Son of York commissioned the writing of the "Beacon" system and a splinter group from "Ease" started the "Apex" system. At the same time, Mr John Brickwood of Lysis Computers in Northallerton wrote a system which he hoped to franchise to various operators on a three-year contract. This was purchased by a consortium of twelve auction companies and became known as LEAN. After about two years it became apparent that nobody was making any profit from any of these.

The scenario was that each auction company would have a computer screen connected to every abattoir who would also have a screen. Each auction company had a highly paid fieldsman whose job it was to persuade farmers who had always sent their fat stock to the mart that this was out of date and the new way was for the fieldsman to grade the stock on the farm to a recognised grid of E to P ("Europ") for confirmation with E being excellent and P being poor. They would also be graded 1 to 5 for fat, with 1 being the leanest. Each mart had a weekly time for their sale and the meat buyers tuned in to bid for the stock on offer. Stock was then sent direct to the highest bidder, thus avoiding the stock coming to the mart.

This often presented problems to the hauliers, who were unable to get articulated lorries into some farms, or found that the lots purchased by an abattoir were located far apart, not conveniently, as in one mart.

Very soon other problems became apparent:

- There were not many bidders.
- There were many disputes as to the grade and, consequently, the money paid to the seller.
- If buyers were short of stock, they would bid an extra penny or two on the system to procure the stock, but then possibly downgrade it to O+ instead of R, causing dissatisfaction for the seller. It rapidly became obvious that the only way to avoid disputes was to send a farmer's stock to the same abattoir on a regular basis and, very quickly, the fieldsmen placed the stock directly rather than through the computer bidding system.
- Many smaller differences were paid by the auctioneer, who could not afford them, and this set a bad precedent which a small number of meat buyers exploited without mercy.
- Many of the buyers also bought in the live auction mart and the auction companies were afraid of falling out with them and losing a bidder at the live auction, as the number of meat companies was already consolidating and there was a great pressure from the supermarket retailers, who wished stock procured direct from the farmers.
- There were also possibly a small number of farmers who may have substituted a part of the stock before they were uplifted. The uplift was often at weekends for the Monday kill and this did not suit many farmers.
- Credit control was difficult, as bidders could not easily be stopped. There was no warning bell to tell the auctioneers that a doubtful client had been buying extra stock from different operators.

Fieldsmen had been introduced directly to mart customers and buyers had been introduced directly to mart customers. Some sellers had found it easier to avoid the auction mart and did not want to lose face with their fellows by having to return to the bustle of the sale ring. The damage to the auctioneers had been done and most firms kept on their fieldsmen. The damage was compounded by mostly paying part of their salary as commission per head thus encouraging these employees to convert more producers away from the auction ring. The damage done to the live prime stock auction system by this technological diversion was major and was, at best, well before its time. The technology of the time, although it appeared wonderful then, was, by today's standards, primitive and completely inadequate.

Whilst the electronic marketing of prime stock was unsuccessful, technology has developed and the internet has become faster thus, in the pedigree world and more so at other sales, the sale can be viewed live by anyone who has the right to log in to view from cameras strategically placed in the ring. With catalogues, pedigrees and photos available on-line purchasers can purchase without attending the sale. This is very common in the fine art world where the auctioneer has a screen beside him showing any remote bids which come in with the buyer's identification. The sale clerk has a key pad on which they constantly change the current bid as the auctioneer moves upwards. This current bid is visible to all online bidders. This system works very smoothly with very few disputes.

The system is also very common in machinery sales where there are often pre-sale viewing days and the sale is then conducted on-line with purchasers from around the world.

In April 2018 Tom Staunton sold eighteen pedigree cattle from his own herd in his on line auction, pedigreesales.co.uk. The cattle were recorded on video and an extended description and pedigree were all available to view on line. Buyers had to pre-register so that their credentials and ability to pay could be checked. The auction was then conducted entirely on line with a top price of 8,400 euros for one of Mr Staunton's own animals from his Caramba Shorthorn herd from Co. Galway.

In Britain, to date, online auctions of livestock not linked to a traditional auction sale have generally been unsuccessful with Farmer Green Auctions, founded in 2015, now closed.

"Sell my livestock" is an online forum where people can list livestock that they wish to sell. Prospective buyers must register and they are then given the seller's contact details to agree a purchase. Payment is processed through a secure website and only paid over to the seller when the transaction is concluded.

Livestock Live is an eBay style operation with a timed auction whereby purchasers can view and bid for stock on online. The seller can accept a bid at any time or can let the auction run until its conclusion. Prices are in Euros and payment is via a secure escrow account.

However with all alternative systems to the live auction, the guide to the value is the auction value. Selling the best stock privately or direct to abattoirs reduces competition in the sale ring with a knock-on effect that puts downward pressure on prices across the board.

All marketing systems have costs to the seller. Those of the auction marts are transparent, those of private and direct sales are often less so although they are there nonetheless.

The electronic age has also enabled the ever-faster production of new regulation and the changing of existing regulation. There appears to be a Tsunami of gold plated bureaucracy affecting producers, hauliers, auctioneers, processors and retailers. Whilst the cost of compliance is small for most individuals the overall cost to the industry is huge in financial terms and can be huge in psychological terms for individuals facing inspections and penalties. Over-regulation hammering the small law-abiding citizen is becoming a major issue. Many feel that regulation was meant to control a problem caused by a minority of irresponsible or criminal actions but that same minority flout the new regulation with impunity leaving the burden on the responsible and law abiding.

CHAPTER XXI
THE LIVESTOCK AUCTIONEERS' ASSOCIATION

The EEC had taken over our agricultural support system in the early 1970`s and the transitional period of five years concluded in 1977 when a flat rate import duty on all sheep meat from outwith the community was fixed at twenty per cent excluding the Irish Republic, which had always enjoyed duty free entry to the British market and I expect they will continue to benefit in 2019 after the UK has left the much changed European Economic Community that we joined forty years ago.

The Common Agricultural Policy has generally been good for our farmers, but Brussels bureaucrats with the over-enthusiastic backing of the British civil service have made many needless obstacles for our farmers to get over and the latter will probably create many more in the future.

The increased speed of communications made auctioneers aware of the need for a strong body to be able to respond immediately to represent their interests, which has, since 1992, been the Livestock Auctioneers Association, which was formed as a replacement for the LAMC with the aims and objectives of the association being:

i) To take all possible steps to develop the system of selling livestock through auction marts, and to enhance the throughput and turnover of markets.

ii) To encourage all markets to be efficient in the conduct of their business.

iii) To operate the highest standards of animal welfare.

iv) To ensure the specialist representation of livestock auctioneers in negotiations and discussions with government, public bodies, national and international bodies and trade organisations.

v) To further the interests of and to raise the profile of auction markets as a method of sale of livestock.

vi) To keep member firms acquainted with legislative and trade developments, which are likely to affect the environment in which livestock markets operate.

The organisation recognised the need for more professional education and in association with the Scottish Institute of Auctioneers and Appraisers, who had been providing some education and examinations for fifty years, they have now established a sound training system in conjunction with Harper Adams University, Shropshire, who provide a four year course supporting some forty students. The teaching staff include experienced auctioneers and valuers under the guidance of Dr Mark Simcock.

The Association has also worked hard to improve the standards of droving staff and Health and Safety issues in marts, while trying, often vainly, to reduce bureaucratic interference. With AHDB a new "Safe Hands" mart drover training programme is being rolled out. The Secretary of the LAA, Chris Dodds, keeps up a constant publicity campaign, combats attacks on auctioneering and leads us positively through the first quarter of the new century.

Presidents and Chairmen of the Livestock Auctioneers' Association

The Livestock Auctioneers' Association – President

1992 – 1995	Stephen D. Clark	Somerset
1995 – 1996	Jim F. Watson	Banbury
1996 – 1999	David J. Tomlinson	Carlisle
	Ian Lawton	Bakewell
1999 – 2002	C. Peter Kingwell	Southern Counties
2002 – 2005	Bruce Daniels	Staffordshire
2006	H. Messer-Bennetts	Cornwall

The Livestock Auctioneers' Association Limited - Chairmen

2007 – 2008	H. Messer-Bennetts	Cornwall
2008 – 2010	A.W. Sneddon	Bakewell
2010 – 2012	Robert C. Whitelock	Hexham
2012 – 2014	Gwyn Williams	Beeston Castle
2014 – 2015	R. Addison	Hexham
2015 – 2016	R.L. Cordingley	York
2017- 2018	J. Brereton	St Asaph
2018 -2019	D Kivell	Exeter

For twenty years Chris Dodds, a Hawick trained auctioneer, has been an excellent and innovative Executive Secretary reporting that, in 2017, for the fourth year in succession the numbers of stock sold at the mart increased with sheep breaking the ten million marker.

In 2017, a conference of auctioneers from nine European countries was organised by the Livestock Auctioneers` Association at Slaley Hall in Northumberland to mark two hundred years of livestock auctioneering, and I was honoured to be asked to contribute to the historical booklet, which recognised the ongoing achievements of the 152 marts then trading in England and Wales.

Over the years, the North of England auction marts have enjoyed immense support from their local newspapers with pages of free but mutually beneficial coverage of their sales. In addition to this, they have the considerable weight of the specialist farming press in the form of the Farmer's Guardian who, through their slogan "the Marts the Heart", give support to many marts. Their large circulation provides a strong platform for advertising and reporting, supported by regular articles by Chris Dodds and auctioneers from around the country.

CHAPTER XXII
THE TWENTY-FIRST CENTURY

In 1998 the sudden closure of one of Britain's most successful marts at Banbury brought home to auctioneers who did not control the freehold and shareholding of their mart the danger of inflated development land values. Often led by supermarket developers looking for a large site, prices of land in country towns had often risen four-fold or more and the owners were keen to cash in. Not only private owners, but also hard-up local councils were keen to get rid of their mart and so a succession of seemingly viable marts were closed, including large centres such as Chelford, Preston, Hellifield, Pannal and Driffield and many historic names have been lost to the business. Some names, of course survive, with their firms concentrating on estate agency as rising prices brought worthwhile fees.

The present century has certainly seen some changes with the giant Aberdeen and Northern Marts, from the North of Scotland, giving up their deadweight business, which included Yorkshire Premium Meat, to return to their roots as livestock auctioneers. Caledonian Marts in Stirling and Longtown Mart have both seen enforced changes of ownership which are fortunately leading to better times for them and, elsewhere, there remains a pattern of merger and co-operation, such as the amalgamation of Kendal and Lancaster Marts, which must be good if it ensures the continuity of the auction system.

In the meat trade, there are constant rumours of amalgamations, some of which have recently come to fruition and thus reduced, again, thc competition in the sector which the marts and farmers need. Credit control remains an issue and auction marts must be constantly vigilant and quick to act. In my young days there was no credit insurance, now it is widespread and, if operated alongside strict control is to be recommended. If you had attended Skipton Cattle Market in the 1850`s and popped in to the White Horse for an ale you would have seen, emblazoned by the landlord above the fireplace, a rhyme which would perhaps also be appropriate for an auction mart office:

Since man to man is so unjust
Man does not know what man to trust
I`ve trusted many to my sorrow
Pay today, I`ll trust tomorrow

At the opening of the century, 2002, following on 2001, was certainly a year of depression among auctioneers as numbers of stock in the marts did not recover easily from a year of Foot and Mouth disease when the marts were closed over a long period with all stock going direct to the slaughterhouses. After re-opening it was discovered that the few supermarkets and their few supporting slaughterhouse companies had decided, unfairly, to boycott the fat cattle rings with numbers in many marts falling by eighty per cent and several marts closing their fat cattle sales completely. With some exceptions, prime cattle producers have not helped themselves by trying to support the marts and they have thus lost the flexibility and quick response to market changes which their fellow sheep producers can still enjoy. Many members of the meat trade look upon the auction marts as the farmer's friend, anxious to keep prices up, and would like to remove them from the field of fair play. This should be resisted by both producers and consumers as the only aim of the multiple shopkeeper is an increase in his profits every year.

Thirsk Calf Show, Matthew Thomson, the author's grandson judging

As the cattle rings have diminished, the meat traders and supermarkets have felt strong enough to introduce irrelevant terms of contract to further harm the marts. They have introduced a so-called residency rule, which brings a heavy reduction in price for cattle, which have been moved several times, in most cases more than three or four moves in the beast's life. Each move to a mart, including the final move to a prime sale may be counted thereby discouraging the sending of calves or young cattle to market; and every time a dealer sells a beast it is one more nail in the coffin and one nearer to a boost in the wholesaler's profit. The auction mart is essential to protect the balance of power in the food chain between the farmer and half a dozen immensely powerful retail groups.

The movement of meat must start with an efficient choice-giving auction mart offering the basic collection of large numbers of stock suitable for all needs and offering buyers of each and every class the chance to buy the specification required at fair prices. Some consumers are already rebelling against the uniformity of taste and weight decreed by the few supermarkets and large abattoirs as these companies aim for an easier life and an easier buck; they like uniform numbers of uniform cuts in uniform packing cases. If the housewife wants choice then the auction system can provide it and so provide for an expanding meat industry.

Meat from the marts has an indisputably good British provenance underpinned by our farmers who produce it to the highest standards of production and welfare found anywhere in the world. Marts have regular inspections and supervisory visits and marts now have safety officers, welfare officers, biosecurity officers and credit control officers, what more do we need? Oh, I forgot about the data-protection officer.

Post 2001, a great help for the marts was the return of cull cows to the ring, which brought extra income to offset the rise in costs caused by new and more stringent regulatory standards and compliance inspections, not to mention the continuing six-day movement restrictions on cattle and sheep (thirteen days in Scotland), which has made life very difficult for many dealers. One essential part of the system was the store cattle dealers, who moved large numbers from every sale and in this century they have practically disappeared as cattle passports and rules have made their existence near to impossible. John Linsley remains as one of the few Northern dealers who still attends our own marts on a regular basis. Trade used to be smoothed out as every dealer had at least a little field or two and could weigh in when things looked cheap. Now the dealer has to think about the movement regulations and dare not buy lest he be timed out from selling stock which is already on hand.

T.B. disease has discouraged farmers, in some areas, from using the marts owing to the work of testing and the fear of buying diseased cattle and to add to the problems, transportation is also being monitored and threatened by some politicians who wish to restrict, or even ban, some movements of live animals.

From 2010 onwards confidence in the auction system has grown in all departments with the exception of the prime cattle ring. Numbers have been rising, credit to prime stock buyers has been more firmly controlled and prices, and thus commission, have tended to rise a little. Costs have, however, spiralled with business rates and effluent charges becoming a big issue. The viability of marts remains threatened by a shortage of profit.

Marts have had to look at new ways of earning a little money to replace the loss of income caused, not least by smaller numbers of prime cattle at their weekly sales and increased costs. Tractors and implements have become a regular feature of the calendar of many marts, often supported by machinery dealers moving on their trade-ins as well as the general flow of surplus farm equipment. Some marts tried selling hay, straw and silage bales but found a tendency for direct sales to bypass the auctioneers for the next load. Two or three sales each year for poultry can be useful with the rarer breeds of bird more profitable than the sale of old hens and dilapidated equipment. Some marts have also developed sales of rare breeds of stock, which, in some cases, have grown into important occasions with a show included, and which can, of course, help to save an endangered breed. Sheep dog sales were all the rage for a few years but in time only two or three sales flourished, in 2019, the Skipton auctioneers achieved a record of 14,000 guineas for a working sheep dog from Emma Gray of Northumberland with averages of £3,223 for bitches and £2,090 for dogs.

Since 1954 there have been many campaigns to encourage more people in Britain to eat more meat, and particularly more lamb, but usually with little success. However, 50 years on, the auction marts have now become more involved in such events as "Love Lamb Week", which is the first week in September, when Rachel Lumley, a north of England sheep farmer, has persuaded many different groups to join together to make a real impact on eating more lamb; many active young ambassadors have visited marts to learn that lamb is "the most natural food" and of the benefits and enjoyment of eating it. Marts encourage visits by schools and youth organisations so that they know of the part they play in the handling of our food and the part they play in raising funds for many essential charitable causes. Many marts are involved in charity fund raising efforts and, for instance, the Lancaster auctioneers recently raised over £20,000 for the North West Air Ambulance and rural benevolent organisations; many other free gift sales have been organised by auctioneers that include many valuable donations and, in helping many good causes, also give publicity to our markets.

As the nation moves through the twenty-first century people must be reminded that the primary role of farming is to feed the rapidly expanding population of the world. Britain must provide as much food as possible and must not be misled by noisy fringe groups who, as online activists or TV pundits, pop up to extol the supposed importance of their opinions on everything from environmental support schemes to telling us, often with no balanced scientific justification, what we should and should not eat. Not only is the message of these people a threat but so is their actual intimidating and inflexible behaviour which is, of course, accentuated by social media.

A pleasant surprise has been the popularity of TV programmes about livestock marts, particularly "The Mart" broadcast from Aberdeen and Northern's fine Thainstone centre, where auctioneers, mart staff, customers, farmers and lookers-on have come over as sensible, salt of the earth people doing a good and worthwhile job showing the essential place of the mart in rural life.

Although old-established, the auction mart industry is not old-fashioned. Innovation is constant, ranging from the embracing of social media for marketing and information purposes to the provision of new types of sales, at new times, including evenings and weekends, and the provision of masses more background information to potential buyers, whether it be Estimated Breeding Values or health status. Internet images and catalogues are common and the mobile phone is now the indispensable canvassing tool.

The marts combine innovation with tradition and good companionship and their auctioneers commit their hearts, their heads and their hands to looking after their customers. When times are hard and friends are few, farmers can depend on the help of the one hundred marts or so that carry on in England and Wales.

APPENDIX 1
LANCASHIRE LIVESTOCK AUCTION MARTS

(See also Chapter 14)

The names given may be those of the operating company or of the auctioneer conducting sales at the time. Dates are often approximate.

Blackburn (Closed 1990)

1874 A new cattle market was erected in Harrison Street

1876 Eli Smith

1880 Whittaker & Armitstead

1890 Blackburn Farmers Mart

1900 William Smith

1930 David R. Hill and Smith

1935 F. Ramsden

Blackshaw Head, Nr. Hebden Bridge (Closed)

1927 Blackshawhead & Farmers Cattle Auction Mart, Shoulder of Mutton Hotel, Walter Taylor

-- Taylor & Taylor of Burnley & Todmorden, auctioneers, monthly cattle sale

Bolton (Closed 1960)

1920 Bolton Famers Auction Co Ltd

1925 John Rodwell, who also held occasional sales at Horwich, six miles away

Bentham

1840 Richard Turner

-- Bentham Auction Mart Limited (new mart 1903)

-- Turner Family

-- Richard Turner & Son, a busy and continuing business

Brockholes

Brockholes Arms Auction Company

1885 J. Kelsall

1927 Brockholes Arms Auction Mart Ltd formed, leased the mart site

-- Joined by Garstang Auction Co.

1948 Purchased the mart site

1968 James Thompson & Son with Geoff Armitstead

1980 I.B. Jackson appointed manager

1985 T. Armistead & Sons, now Armitstead Barnett

2000 Richard Furnivall of Armitstead Barnett

-- Brockholes Arms Auction Mart continues enthusiastically today, it benefitted from the closure of Preston Mart.

Burnley (Closed 1920)

1886 New Mart built, Ackroyd and Green

1907 Fortnightly sales held by Herbert Lee

Carnforth (Closed 1967)

1870 Mr Parkinson monthly sales in Queens Hotel Yard

1954 I.B. Jackson moved to Brockholes. Carnforth acquired by Kendal Auction Company

By 1960 weekly sales of 200 to 300 fat sheep were held on Wednesdays and a few store lambs and ewes in the autumn. Did not re-open after 1967 Foot and Mouth outbreak.

Chipping (Closed 1955)

1901 The Clitheroe Auction Mart Company Ltd

Clitheroe

1885 Samuel Green

1897 The Clitheroe Auction Mart Company Ltd

Silverwood & Balshaw

1988 New Mart built

1995 The Clitheroe Auction Mart Co. acquired Silverwood &

Balshaw, Joe Balshaw`s cattle sales incorporated into the Clitheroe business

-- Continues successfully as The Clitheroe Auction Mart Co. with Silverwoods as their fine art division.

Cliviger, Nr. Burnley (Closed)

1925 Holme & Cliviger Auction Mart, Walter Taylor of Cliviger & Todmorden, auctioneer

Colne (Closed 1957)

1674 "Colne hath a good market on Wednesdays for cattle & provisions"

1868 Monthly cattle market & quarterly fairs were held

1898 Problems being raised with injuries & congestion at the market

1900 "Good weekly numbers" were being recorded but no actual numbers, so possibly they were not so good

1905 Local Authority erect a new auction mart at Ludgate Circus, traditional market and fairs ceased

-- Initially, fortnightly sales on a Friday with 60/100 cattle plus sheep & pigs

-- Various auctioneers sold here but none appear to have been successful for very long

Garstang (Closed 1927)

1880 Henry Thornton – Monthly sales

1890 Garstang Auction Mart Company – Weekly sales

1927 Moved to Brockholes Arms Auction Mart

Gisburn

1911 Lord Ribblesdale

1922 Gisburn Auction Mart Limited

1952 Richard Turner & Son, Auctioneers and Secretaries

-- Continues today as a well run operation

Haslingden (Closed 1986)

1890 James Barlow and Son

1920 Haslingden Cattle Market Co Ltd

1950 Norman Duxbury, Manager

1960 Albert Mawson

1970 Joe Worthington

1980 Andrew Birtwhistle

Cockerham, Nr. Forton (Closed)

1890 Tom Armitstead took over from George Corbishly

Horton in Ribblesdale (Closed 1930s)

1933 Richard Turner & Son, store cattle & sheep at the Golden Lion Hotel

Kirkham (Closed)

1880 Topham & Son (every Monday)

1890 E.G. Hothersall (Every Tuesday in Railway Hotel yard)

Lancaster

Cattle Market was in Thurnham Street

1880 G. Corbishly (on Wednesday)

1880 W. Haythornthwaite (on Tuesday)

1890 James Downham

1890 Thomas Armitstead bought Corbishly`s business

1900 Lancaster Farmers Auction Mart Company Ltd

1920 Reg. & Herbert Armitstead, T. Armitstead & Sons

1940 Jacky Johnson partner with Thomas J. Armitstead

-- Succeeded by Geoff Armitstead

1965 New mart erected

2005 Amalgamate with Kendal Mart & M.B Hodgson & Son

2013 North West Auctions Ltd (company formed). New mart erected at M6 junction 36 to replace Kendal, Lancaster remains busy

Liverpool (Closed 1960)

See Chapter 5

Longridge (Closed 1939)

1880 Henry Holden in Corporation Arms Yard

1920 E.G. Hothersall & Sons

Manchester & Salford (Closed 1964)

See Chapter 5

Oldham (Closed)

1804 Weekly market held in Old Market Place, but it became a Police Burgh in 1834 when all livestock were banned from the town and sales never recovered. Thomas Whitaker, auctioneer, tried much later but got no support.

Pilling (Closed 1980)

1880 John Singleton at Stalmine Mart

1890 Pilling Farmers Auction Company Ltd

1945 T.J. Armitstead & Geoff Armitstead (T. Armiststead & Sons)

Poulton-le-Fylde (Closed 1960)

1900 E.G. Hothersall & Sons

1954 Ben Johnson, auctioneer

At the first weekly sale in September 1900, 726 fat sheep, 45 fat cattle, 51 pigs and calves were sold with bullocks topping at £15. The import of Irish & Scottish cattle through Fleetwood enabled a fortnightly cattle market to be held at Poulton

Preston (Closed 2001)

1860 Farmers Auction Mart Limited

1867 E.G. Hothersall & Sons

1983 Joined with Kendal Auction Company

Rawtenstall and Bacup (Closed 1979)

1920 Edward Ashworth, followed by sons and grandsons

1939- 1954 Not a Collecting Centre – Dairy Cattle were sold weekly

Rochdale (Closed 1951)

1872 Horse Fair held in Cheetham Street

1887 Three Cattle fairs per year held until this date

1911 Cattle Market in Manchester Road

-- E. Robinson & Sons

1951 Cattle Market at The Holme discontinued

St Helens (Closed 1955)

-- Mart on Carr Mill Road

Warrington (Closed 1960)

1880 Sales held in the yards of two inns, The Eagle & Child and The Lion

1885 W. Bainbridge

1890 John V Mainwaring at Eagle and Child Yard, weekly sale

1890 Andrew Hutchinson of Knutsford at The Norton Arms, weekly sale

1895 New Corporation Mart opened at the Old Militia Stores

1900 W Bainbridge, Froghall Lane

1900 Lancashire and Cheshire Farmers Association with Hutchinson as auctioneer

1930 J Mainwaring, Stafford & Co (principal, Stanley Wright)

1940 Marsh & Son, Nantwich, founded 1875, & selling at the Norton Arms Cattle Market

Whalley, Nr. Clitheroe (Closed, probably prior to World War I)

1910 Martin Cooke, fortnightly cattle sales on a Thursday

Wigan (Closed 1992)

1866 Mr Walker commenced small weekly sales of livestock at the Ship Hotel

1896 Green and Trickett

1901 John Trickett bought the business

1908 Part of Ship Yard added to the mart

1925 A son, Oswald Trickett FAI (by exam), joined the firm followed by his brother, Andrew

1937 New mart built in Prescott Street

1964 Tom Goulden joined the firm (married Miss Trickett)

1966 Oswald Trickett died

1969 Andrew Trickett died

1992 Tony Clough and Tom Goulden sold the mart for building

APPENDIX 2
YORKSHIRE LIVESTOCK AUCTION MARTS

(See also Chapter 15)

The names given may be those of the operating company or of the auctioneer conducting sales at the time. Dates are often approximate.

Apperley Bridge, between Bradford and Leeds (Closed 1940)

1870 Thomas Burroughs (Queens Hotel Yard)
1880 W H Pountney
1898 Mason & Jennings
1900 Apperley Bridge Livestock Auction Co Ltd

Barnsley (Closed 1990)

1845 Mr Wilby held sales in Denby Dale
1860 Wilby's in new Barnsley Corporation Market
1890 Lancaster & Sons also held sales
1933 New Corporation Market built with Wilby's continuing to sell
1935 C.E. Smith & Co started weekly sales
1961 Wilby & Son, continuing selling weekly

Bedale (Closed 1990)

1880 Various auctioneers selling in the town`s market
1890 Bedale Auction Co
1920 J. Hopper and Son, Leyburn
1935 Oswald Lister
1960 Taylor and Wetherall
1971 Norman Brown & Gordon Pratt

Beverley (Closed 1998)

1891 Ullyots
1930 R Hornsey & Son
H Jackson & Son

Leonard Nutchey
Clegg & Wright
Dee & Atkinson
Hill & Leonard
M Featherston
Evinson & Son
FW Ullyot & Son

Bingley (Closed 1996)

1890 Tillotson & Son
1906 William Ash of Garforth acquired lease of mart
1914 Bingley Livestock Auction Mart Co. Ltd
1920 They purchase the freehold
1922 Fred Lister appointed auctioneer
1930 New pens and sidings erected
1953 Otley Cattle Market Co. Ltd purchase Bingley Mart
-- William Lister, manager
1982 Chris Clubley, cattle auctioneer
1993 Martin Lister, chief auctioneer
(See Chapter 13, the Listers)

Birstwith

Oswald Lister's office, there was no mart here
He sold at Bedale and Otley
His son Michael sold at Wharfedale from c1960-1995

Boroughbridge (Closed 1990)

-- Home to Barnaby Fair, one of the largest fairs in Yorkshire
1910 Boroughbridge Farmers Auction Mart Ltd (capital £15,000) sold on the Market Hill
1920 Basil Johnston
1930 Basil Johnston
1936 James Johnston
1985 Paul Johnston

1988 Sold to Thirsk Farmers Auction Mart for £540,000

Bradford (Closed 1965)

1852 J. Wright

1871 New Cattle Market, Abbatoir & Wholesale Meat Market erected by the council

1900 Fred Deighton holding weekly sales

1910 Murgatroyd Smith

1935 Tordoff

Bridlington (Closed 1970)

1900 Hornby Cranswick

1920 Cranswick & Cranswick

Burley in Wharfedale (Closed 1915)

In a report to the Highland and Agricultural Society it said that sales of cattle there were aided by one of the first recorded cattle weighbridges.

Brighouse (Closed 1939)

1896 Davison and Sunderland held monthly auction sales. On 13th October 1900 they sold horses, cattle, sheep, pigs and a goose fair.

Burstwick (Closed 1939)

1920 R. Leonard & Son

Situated nine miles east of Hull, monthly sales of sheep, store cattle and pigs were held by R. Leonard and Son of Hull until 1939. It was not approved as a wartime Collecting Centre

Castleton

-- Seasonal site. Annual Castleton Sheep Sale held by Richardson & Smith, ceased 2001, now held in Ruswarp Mart.

Catterick (Closed 1930)

At one time sales were held beside the Bridge Inn

1900 Catterick Station Farmers Auction Mart Co. Ltd, Brompton on Swale

-- C. Layfield, company secretary

1910 P. R. Harland, auctioneer

-- Gordon and Cecil Pullen, auctioneers

Dunswell – Between Hull and Beverley

2002 Frank Hill & Son

-- Continues today, trading as "Hull Cattle Market"

(See Chapter 5, Hull)

Doncaster (Closed 2001)

1870 T. R. Miller

1880 Thomas Shearman

1935 Shearman & Johnstone

1955 Jennings & Pickup

Sidall & Reynolds

Hawson, Merrywether & Corbett

Bendall & Co

1970 All remaining firms joined together as Doncaster Market Auctions Ltd

Driffield (Closed 2001)

1860 George Clark

1880 Francis Cook

1900 Hornby & Wilberfoss

5 Firms operated incl. Messrs Jarrett who owned the Cattle Market

1910 Cook & Dee

1935 Cranswicks and Ullyot & Sons also sold

1950 Dee & Atkinson

C.B. Shipley & Featherstone

1980 Dee & Atkinson, Cranswick & Ullyot - Joint Market Co formed

Easingwold (Closed 1980)

1890 R. Burton (12 cattle)

1935 R. Burton had weekly sales of 80 cattle, 300 sheep and 120 pigs
1956 Frank Close & Simpson
1960 H.D. Robinson & Son

Fadmoor

1920 Cundalls held seasonal sales
-- Seasonal site, Cundalls hold annual sales

Farndale

-- Seasonal site, Chris Clubley hold annual sales

Fewston (Closed 1961)

1900 F. Lister, seasonal sales
1935 F. M. Lister sold 2,460 store lambs & 22 cattle on 1st October
-- When Listers started holding 5 Autumn store sheep sales at Otley, clients were encouraged to go there
1961 Sales moved to Lister's Otley Mart

Glaisdale

-- Seasonal site with sales by Richardson & Smith, ceased 2001

Goathland (Closed 2000)

1900 Robert Harrison held his 30th Annual Sale of 1,800 sheep
An established sheep mart latterly conducted by Boulton and Cooper.

Goole (Closed 2001)

1890 Goole Auction Marts
1935 Clegg & Son

Guisborough (Closed 1983)

1900 Cleveland Auction Co Ltd
1930 W. Hedley, Auctioneers
1980 Join with Stokesley Mart

Halifax (Closed 1965)

1890 Davis, Son & Sunderland

1900 Ackroyd & Son

1915 Braithwaites

1930 Henry Spencer

1935 Irish fat cows arrived every week, making up to £15

-- It continued to attract business until 1965.

Hawes

1887 Hawes Auction Mart (T T Iveson)

1914 Hawes Farmers Auction Mart Co Ltd

1960 William Preston, Manager

-- Under Raymond Lund it remains one of the leading sheep marts in England

Hellifield (Closed 1990)

1870 Hellifield Auction Mart Co (Messrs Iveson)

1950 Malcolm Skidmore

1986 Craven Cattle Markets, Skipton

Helmsley (Closed 1985)

1875 Monthly Sales

1885 Weekly Sales

1889 Seth Kirby

1921 Seth Kirby & Sons

Holmfirth

-- Holmfirth Mart Co

-- A.E, Wilby & Son

1866 Wm Sykes & Son

-- Herbert Sykes

`` William Sykes Shaw (nephew of Herbert)

1951 Vernon Dixon joined Wm. Sykes & Son as auctioneer

1951 Holmfirth Attested Auction Mart Ltd, set up in opposition to existing mart

1959 Vernon Dixon took over William Sykes & Son, but Sykes ceased selling in Holmfirth

1980s New sale ring & extensive improvements at the Attested Mart

1983 Paul Dixon (son of Vernon) of Wm. Sykes & Son selling at Huddersfield & Penistone

2008 Paul Dixon sheep auctioneer, David Moxon cattle auctioneer

2018 The Holmfirth Attested Auction Mart continues to thrive, Anthony Hobson, general manager. Some autumn sheep sales held in conjunction with Wm. Sykes & Son

Hope (Closed 2001)

1900 Siddal & Reynolds, Sheffield

1980 Bagshaws

Howden (Closed 1920)

1870 Mr Pratt (Wellington Hotel Yard)

1900 James Glew

1910 J.F. Palframan

Huddersfield (Closed 1996)

1875 Wm Sykes & Son

Hull (Closed 1995)

See Chapter 5

Ilkley (Closed 1931)

1906 Thomas Hartley

-- Dacre, Son & Hartley

1920 F. Lister

Ingleton (Closed 1954)

1880 R. Turner & Son

Keighley (Closed 1939)

1905 Ackroyd & Son

Knaresborough (Closed 2001)

1890 Joseph Horner & Co.
1909 George Thornton
1920 Tony Thornton
1970 Thornton & Linley

Kirkbymoorside (Closed 1973)

1887 Bartliff & Dale
1930 Tom Kendrew
1954 Kendrew & Dennis
1968 Joseph Cundall & Sons, Peter Woodall, manager

Leeds (Closed 1954)

1870 Lodge & Bartle
1875 W.H. Pounteney
1890 Heppers
See Chapter 5

Leyburn

1870 John Atkinson
1894 Joseph Hopper (established 1886)
1918 Leyburn Livestock Auction Mart Co
-- Continues successfully, Stephen Walker, manager

Long Preston (Closed 1972)

1900 Thomas Preston
1930 Robert Preston
1959 James Thompson, Kirkby Lonsdale

Lythe (Closed 1920s)

-- Alan Richardson
1920s Merged with Smiths of Whitby to form Richardson & Smith, mart closed to focus on Ruswarp

Malton (1)

Mart shared over the years principally by Boulton & Cooper & Cundalls but some other firms, including Dee & Atkinson sold there for a time

1830 James Thompson & Son

1875 Robert Boulton

-- Boulton & Constable

1881 Boulton & Cooper

1987 Join with Stephensons of York

Malton (2)

1878 Botterill & Sons

1885 Cundall & Son (Established 1860)

-- Joseph Cundall & Sons

1982 Wells Cundall

-- Cundalls now owned by the Watson family

-- Mart continues, operated as a joint arrangement with Stephensons, as Malton Livestock Auctioneers

Masham (Closed 2007)

1887 R. Imeson

1900 Robert Imeson and Son

1921 Farmers Mart Co Ltd

I.W. Clements

1935 R.F. Broadley

1940 Don Johnston, Great Hammerton

-- Masham Farmers Livestock Mart Co.

2002 Craven Cattle Marts, Skipton

Meltham (Closed 2000)

-- Seasonal Sheep Fair conducted by William Sykes & Son

-- Meltham sheep fair continues to be conducted by William Sykes & Son & the Holmfirth Auction Mart but now at Holmfirth Mart

Market Weighton (Closed 1914)

1880 Frequent sales, mostly store stock, were held by R.M. English & Son

Northallerton (Malpas Mart)

1876 Mr Malpas

1910 North Yorkshire Stock Market Co (Harker and Farrow Auctioneers)

1954 Taken over by Applegarth Mart

Northallerton (Applegarth Mart)

1887 John Todd`s Northallerton Store Stock Market

1907 Applegarth Mart (John Todd)

1912 Northallerton Auctions Ltd

1947 Darwin, Metcalfe, Alsop & Pratt were auctioneers

-- Continues successfully today, Giles Drew, mart manager

Northallerton (Station Mart)

1875 Robert Atkinson

1960 Thomas Hedley

1975 Taken over by Applegarth Mart

Otley (1, Listers) (Closed 2000)

1880 Horner

1890 Joe & Fred Lister

1922 Joined with Wharfedale Farmers Mart Ltd.

1931 Otley Cattle Market Auction Ltd, William Lister, set up in opposition

1935 Bridge End Mart opened by Otley Cattle Market Auction Ltd.

1950 Selling TB Attested cattle only, very successful dairy sales

1952 Mervyn Lister

1978 Martin Lister

1990 Graham Lister

2000 Bridge End Mart closed

See Chapter 13, the Listers

Otley – (2, Wharfedale)

1870 Dacre & Son

1893 Wharfedale Farmers Auction Mart Ltd., East Chevin Road Mart

1931 Joined by J. & F. Lister

1962 Ben Atkinson, auctioneer joined from Lister`s, later company secretary

1980s 5/600 Prime cattle, 1,000/1,500 sheep, 400 pigs & 400 calves

-- A successful business today, Ian Smith, manager, Ruth Priestley, company secretary

Pannal (Closed 1994)

1935 R. Burton of Easingwold

1954 H.D. Robinson & Sons

From 1960 to circa 1980 this was a most successful mart, among the ten largest in Yorkshire.

Pateley Bridge

1900 Pateley Bridge Cattle Market

1930 Oswald Lister

1951 Atkison & Worthey

1970 Worthey & Smith

2009 Took over Sawley Auction Co.

-- Continues operations under enthusiastic management

Penistone (Closed 2004)

1880 Wm Sykes & Son

1900 Wilby & Son

1954 Siddall & Reynolds

1960 C.E. Smith & Son

1983 Paul Dixon of Wm. Sykes & Son

-- With Wilby`s on a Monday

-- With Bagshaw`s on a Thursday

2004 Barnsley Council sold mart to Tesco. Every pen full for last sale.

Pickering (Closed 1980)

1910 H. Creaser & Son

1920 Boulton & Cooper

Pocklington (Closed 2000)

1920 R.M. English & Son (store stock sales)

1960 Chris Clubley & Co

Pontefract (Closed 1965)

1863 Lowther & Son

1870 Mr Dodds

1895 Bentley & Sons, later bought by Sykes & Son

1939 Taylor & Co

-- Bendall & Co, September 1964 reporting 46 cattle, 60 sheep & 145 pigs

Ravensworth, Nr. Scotch Corner (Closed 1932)

1895 John Richard Deacon held monthly store stock sales at Fox Hall

.

Reeth (Closed 1991)

-- Seasonal site

-- Addison family, Leyburn Auction Mart & F H Brown

Richmond (Closed 1990)

-- Sales held in the town square

1920 Richmond Farmers Mart Ltd

-- Addisons, auctioneers

1987 Stokesley Auction Co.

Ripon (Closed 2000)

1890 Ripon Farmers Mart Ltd

1935 Seth Kirby & Sons

-- Tom Kirby

-- R. H. James appointed managing auctioneer

Rotherham (Closed 1967)

1850 September -365 cattle and 3,340 sheep were in the weekly market

1881 Styring & Taylor

1890 Merryweather & Corbett

1926 New mart built in Corporation Street

1935 Shearman & Johnstone

1954 Merryweathers succeeded by John Dawson

1954 The two firms joined to operate the market together

Ruswarp (Nr. Whitby)

1902 Whitby & District Farmers & Butchers Auction Co Ltd

1920 Richardson & Smith (Moved from Lythe, which closed)

-- Continues to thrive

Sawley (Closed 2005)

1948 Clarence Nichol traded as Sawley & District Farmers Mart

1953 Joined with Pateley Bridge Mart

Settle (Closed 1940)

1895 J. Kay

Spacey Houses, near Pannal (Closed)

1893 R. Burton

1900 Mr Robert Burton sold without reserves on Monday 7th September: 172 cattle and calves, 1,023 sheep and lambs & 302 pigs

1914 R. Burton possibly moved to Pannal

Seamer Junction (Closed 1992)

1880 Boulton & Cooper

Selby

1873 to 1986 Bartle & Son

1890 Richardson & Pearce Brown

1906 Richardson & Smith

1960 Selby Auction Mart Co

1987 Moved to new mart

-- Bartle family remain the largest shareholders and still going strong

Sheffield (Closed 1960)

1870 J. Bannister

1890 Styring & Taylor

1930 Siddall & Reynolds

Skipton

1880 Mr Pallister

1890 Skipton Auction Mart Co

1975 Craven Cattle Marts Ltd

1984 New mart built

-- Continues as a major centre

Stokesley (Closed 2004)

-- George Hodgson

1890 Holmes & Ross

1900 Hodgson and Farrow

1900 William Hedley

-- Two marts, Station Mart & Town Mart

1905 Stokesley Butchers Mart Ltd

1905 Stokesley Farmers Auction Co Ltd

1935 H. Richardson & R. Dobson, joint auctioneers

1965 Two marts amalgamated and one mart sold

1980 Cleveland and Stokesley Auction Companies almagamated

1992 Acklington Auction Co

1995 Bought by Penrith Farmers & Kidds, Penrith

Tadcaster (Closed 1990)

1869 Thomlinson & Sons

1871 Stephenson
-- Thomlinsons until 1990

Thirsk

1875 George Freeman
1900 J. H. Freeman
1907 Thirsk Farmers Auction Mart Ltd. open in opposition
1918 Freeman`s closed
1965 Seth Kirby & Sons
1972 Thirsk Farmers Auction Mart Ltd
2006 New mart built, continues as a busy centre

Wadsley Bridge (Closed 1960)

1930 Siddall & Reynolds, Victoria Arches Mart, Sheffield

Wakefield (Closed 1965)

1870 Mr Dodds
1887 Dodds & Robinson
1960 M. Thomlinson & Sons
See Chapter 5

Wetherby (Closed 2001)

1875 Sergeant & Son
1880 M. Thomlinson & Sons

Whitkirk – Near Leeds (Closed 2001)

1890 Bartle & Son
See Chapter 5

Wombleton

-- Temporary pennage. Cundalls hold seasonal sheep sales

Yarm (Closed 1970)

1890 Robert Hedley, who also sold at Cleveland, Guisborough & Stokesley

1919 R. & W. Hedley (Robert & William Hedley)
They also held annual sales of seed potatoes in Darlington
1935 William Hedley
1970 Hedley & Son (Closed)

York

1868 Hands & Richardson of Caistor
1870 Mr Walker (Spotted Cow Yard)
Mr Dodds from Wakefield
1880 Thomlinson & Sons (ceased in 1968)
1889 Stephenson & Son
1890 Ross & Co
1900 R. M. English & Son
1920 Joseph Cundall & Sons
-- Don Johnston
1971 New Mart at Murton (Joint operation by Stephenson and English)
-- York Auction Centre remains a busy & diversified mart

See Chapter 5

APPENDIX 3
GUARANTEED PRICES FOR FAT STOCK DEFICIENCY PAYMENT SYSTEM 1948

A Memorandum prepared by the Joint English and Scottish Livestock Auctioneers Consultative Committee

1. A system of deficiency payments paid up on fat stock sold either by auction or at a deadweight certification centre, at the producer's option, and administered by the Livestock Commission, is put forward as the best practical means of giving effect to the Government's price guarantee for fat stock, given in the Agriculture Act, 1947, after the end of the present control, and as the best practical long-term policy for ensuring the stability and prosperity of the livestock industry, and at the same time allowing the maximum freedom to all sections of that industry.

2. In the following description of the proposed system, there is no discussion of the way in which the fund required for making deficiency payments should be provided. Obvious alternatives are to draw this fund either from the general revenue of the country, or from a levy on imported meat, or from a levy on both imported and home produced meat. It makes no difference to the operation of the following scheme which way is adopted for raising such deficiency payments as may be necessary; and as this is a matter of financial policy which has no bearing on marketing methods, no further reference is made to it here.

3. The deficiency payment system is one which permits an immediate and smooth change-over from the present full control to an open market system, yet still gives the producer guaranteed prices for fat stock.

4. The Livestock Industry Act, 1937, and the Agricultural Development Act, 1939, were on the Statute Book before the war, and the Agriculture Act, 1947, has since been passed. If additional powers are required to bring the scheme here proposed into force, the necessary new legislation would only be the development of principles already recognised.

5. The deficiency payments would be related to the prices fixed from

time to time in accordance with the Government's guarantee under the Agriculture Act, 1947.

Principles applying to all Fat stock

6. Fat stock would be presented at authorized Livestock Certification Centres or Dead Weight Certification Centres at the option of the producer. Advance notice of entry would have to be made in each case. This would prevent last minute entries from flooding the market, and would enable auctioneers to notify purchasers weekly of the stock coming forward and so improve market intelligence.

7. At Livestock Certification Centres stock which had been properly entered would be examined by the Certifying Panel who would decide whether stock qualified for deficiency payment or not, and in what classes the qualifying stock should be placed. If found suitable for slaughter, and then sold the stock would qualify for deficiency payments in accordance with the regulations in force at the time. No stock would be eligible for certification unless it had been entered as provided in the foregoing paragraph; but provision should be made for late entry in exceptional circumstances. What classes of stock should be certified as suitable for deficiency payments is a matter to be decided by the Livestock Commission in consultation with producers.

8. No grading would be necessary, as the competition by the purchasers, operating in an open market, would regulate the value of stock according to class and quality. This would encourage the efficient production of high quality stock. The producer would receive the price of his stock arrived at by open competition plus the deficiency payment. To have a number of difference grades would only complicate the working of the scheme and involve delays.

9. After the classification by the Certifying Panel, stock would be sold at the auction, the purchasers being entirely free to choose their purchases and to bid accordingly.

10. After sale, stock certified for deficiency payment and sold would be marked in accordance with the regulation laid down by the Livestock Commission.

11. No purchaser would be allowed to remove the stock from the centre until such marking had been completed.

12. The Certifying Officer would make out and forward to the Livestock Commission a Deficiency Payment Certificate in respect of each lot

of stock certified and sold. These certificates would show in the case of cattle and pigs the net live weight and the actual price realized on sale. If the Commission wished to check the classification of stock by following up individual animals after leaving the Certification Centre, it would help it in its task to have the name of the purchaser entered on every certificate.

13. The flow of stock to the markets could be controlled by varying the guaranteed price from one part of the year to another. Guaranteed prices for the earlier months of the year should be so fixed as to encourage an even flow of stock to the markets throughout the year; and the guaranteed prices applicable to each year should be announced early enough to enable producers to plan ahead.

14. For the purpose of the deficiency payment ascertainment, the computation period for working out the average price should be one of four weeks at a time. The Livestock Commission would receive all certificates issued for stock certified and sol during each four-week period and would make its computation on the completion of each such period. Payments by the Livestock Commission to the producers might be made quarterly.

APPENDIX 4

The following is a list of weekly sales in the three counties in the latter part of August 2017, which shows the great variation between marts and warns us that, at some centres, numbers are getting dangerously low but, of course, seasonal variations come into play.

	Fat Lambs	Ewes	Store Sheep	Cull Cows	Prime Cattle
Beeston	459	85	500	115	38
Bentham	2,815	1,479	5,360	13	
Brockholes	643	222	455	44	7
Clitheroe	1,055	328	1,455	13	10
Gisburn	1,607	120	3,600	116	41
Hawes	1,087	80	12,000	20	
Holmfirth	336	173	400	15	5
Hull	96	30	125	5	
Lancaster	666	188	2,000	74	15
Leyburn	868	394	1,665	18	3
Malton	652	168	6,620	9	52
Northallerton	798	148	1,880	78	150
Otley	804	210	1,300	35	17
Ruswarp	140	153	1,200	29	2
Selby	419	130	420	26	319
Skipton	2,956	667	9,232	46	15
Thirsk	1,227	152	900	54	217
York	510	102	1,000	5	112

APPENDIX 5

TESTIMONIAL
presented to
R.S. HARRISON, Esq., F.A.I.
by the
COUNTY AND DISTRICT CHAIRMEN OF AUCTIONEERS
in appreciated of his Valuable Services
as
Area Livestock Supervisor
for the
NORTH-WESTERN AREA
1939 – 1942

CHESHIRE

County Chairman

J. WRIGHT, Esq., F.A.I.	Alpraham, Tarporley

District Chairmen

J.W. WRIGHT, Esq., F.A.I.	Beeston & Wrenbury
L. ARNOLD, Esq., A.A.I.	Altrincham
F.R. ARNOLD, Esq.	Chelford
W.J. WHITTAKER, Esq.	Congleton
E.E. WRIGHT, Esq., F.A.I.	Crewe
W.H. BRADY, Esq., P.A.S.I., A.A.I.	Marple
A.S. TURNER, Esq., F.A.I.	Macclesfield

LANCASHIRE – NORTH

County Chairman

JOHN HOTHERSALL, Esq., F.A.I.	Preston

District Chairmen

J. HOTHERSALL, Junr., Esq.	Preston
JOE HOTHERSALL, Esq.	Poulton
J.C. COWARD, Esq., F.A.L.P.A.	Ulverston
J. HALSTEAD, Esq.	Bolton

J.E. BARLOW, Esq.	Haslingden
J. TRICKETT, Esq., F.A.I.	Wigan
R. CLARKSON, Esq.	Pilling
J. JOHNSON, Esq.	Colne
D. SMITH, Esq., J.P.	Blackburn
R.P. KIRKBY, Esq.	Lancaster
E. CHESTER, Esq., F.A.A.	Clitheroe
T.B. JACKSON, Esq., F.N.A.A.	Claughton

LANCASHIRE – SOUTH

County Chairman	
G.H. TOMMIS, Esq., F.S.I., F.A.I.	Stockport
District Chairmen	
G.S. MARSH, Esq., F.A.I.	Warrington
S. WRIGHT, Esq.	St. Helens
E. ROBINSHAW, Esq.	Rochdale
G. COOKSON, Esq.	Manchester
J.A. JACKSON, Esq.	Liverpool
Liverpool Officials	
Capt. A.D. HARPER	Liverpool
N. SHRINKFIELD, Esq.	Liverpool
A. WOOD, Esq.	Liverpool
T.G. FORD, Esq.	Liverpool

WESTMORLAND AND PENRITH

County Chairman	
Capt. A.L. KIDD, F.A.I.	Penrith
District Chairmen	
J.W. PROCTER, Esq.	Penrith
J.H. GREENFIELD, Esq.	Kendal
E. HUCK, Esq.	Kirkby Stephen
R. TURNER, Esq., F.A.I.	Kirkby Lonsdale
J.R. LONGRIGG, Esq.	Lazonby
T. SISSON, Esq.	Appleby

APPENDIX 6

SCOTTISH LIVE STOCK AUCTIONEERS' AND SALESMEN'S ASSOCIATION.

REPORT BY SECRETARY presented to the Annual Meeting of the Association, held on 4th December, 1919.

1919.

W. HUTCHISON, LL.B.,
Secretary and Treasurer,
223 St. Vincent Street, Glasgow.

R. E. Robertson, Ltd., 198 Buchanan Street, Glasgow.

SCOTTISH LIVE STOCK AUCTIONEERS' AND SALESMEN'S ASSOCIATION.

REPORT BY THE SECRETARY presented to the Annual Meeting of the Association, held on 4th December, 1919.

In presenting my report of the second year's work of this Association, I feel that I must of necessity limit my record to the more important features, as to dwell in detail, in the many points considered and dealt with by the Executive Committee, would only present you with facts already known. The Executive Committee have had eleven meetings. In addition, there have been three Sub-Committee meetings, and two meetings of the Association, and the most important matters considered, and which I propose to refer to, are: Surplus Stock, Super-Grade Prices, Increased Allowances for Clerical Work, Simplification of Returns, and Increase of Commission. After considerable negotiations, extending over a lengthened period, and embracing numerous interviews, the Executive Committee were able to come to a definite Agreement with the Ministry regarding Surplus Stock, and got an Agreement which was sanctioned by a Special General Meeting of the Association, drawn up and signed, and which Agreement had, in addition, embodied in it the full terms of Agreement between the Ministry and the Association. Several firms, unfortunately, did not seem to clearly understand what Surplus Stock actually was, and the returns to me in many cases were so inaccurate that great delay was caused in obtaining settlement of what was due in respect of Surplus Stock, and I would hope that the exact position is now clearly understood, and that *the necessary returns will be sent me regularly and accuratety, to enable me to check promptly and efficiently the figures given me with those submitted to me by the Ministry.* In the general interests of the members this is essential. In respect of all stock handled, whether graded animals or D.W.B., returns should also be regularly made to me. Delay and inaccuracies prevent prompt payments of what is due to the Association. Allowances were obtained in respect of the clerical work on the basis of the stock handled, and it is hoped that these will, to a certain extent at least, compensate members for the great amount of clerical work involved in making up the various returns. A Sub-Committee of the Executive Committee was able to get the returns greatly simplified from what was originally proposed, At a Special General Meeting of the Association it was resolved to forward a request for increased commission, a deputation being appointed to interview Mr. MacDougall on the matter. This Committee duly met Mr. MacDougall, who listened carefully and sympathetically to the request, and promised to do his utmost to get something done. They also had a meeting with the Live Stock Accountant, who agreed to recommend an increase. As a result a member of the Costings Committee of the Ministry from London came North and inspected the books of a number of firms, but, unfortunately, I am afraid his report was not as favourable as it might have been, and, in consequence, nothing in the way of any increase has yet been granted. Owing to the discrepancies existing between the prices of graded cattle and those payable in respect of cattle slaughtered on the D.W.B., during the spring of last year the Executive Committee felt that something must be done to equalise the prices or to get increased prices for graded cattle. As no definite result could be obtained from the correspondence which took place, it was resolved to send a deputation to London to interview the Ministry on the matter, and they proceeded to London on March 4th, 1919. A deputation from the Farmers' Union had been in London on the same matter the previous week, and had been unable to obtain any concession; but I am glad to be able to record that the facts and figures adduced by the members of the deputation had the desired effect, and obtained what was known as super-grade price for all grades of cattle, and got the same into operation almost immediately. Unfortunately, this was withdrawn in the beginning of June, and practically the same deputation again proceeded to London to urge the renewal of the concession, but, unfortunately, they were unsuccessful, though they were able to impress on the Ministry the necessity of equalising the prices that they should pay in respect of graded stock and those on the D.W.B., and the new schedule of prices, I understand, bears evidence of this.

I would again express my appreciation of the courtesy and consideration extended to me by the members generally.

W. HUTCHISON,
Secretary.